Lecture Notes in Mathematics

A collection of informal reports and seminars
Edited by A. Dold, Heidelberg and B. Eckmann, Zürich

36

Armand Borel
The Institute for Advanced Study Princeton, New Jersey, USA

Topics in the Homology Theory of Fibre Bundles

Lectures given at the University of Chicago, 1954
Notes by Edward Halpern

1967

Springer-Verlag · Berlin · Heidelberg · New York

Library of Congress Catalog Card Number 67-28561 . Title No. 7356.

INTRODUCTION

This fascicle consists of the Notes of a course given at the University of Chicago in 1954, the purpose of which was to discuss some then recent developments in the homology theory of fibre bundles, pertaining to H-spaces, spectral sequences, classifying spaces and characteristic classes. Since then, of course, alternative approaches to some of these topics have been introduced, and new results have been obtained, which make these Notes outdated in several respects. In spite (or maybe because) of this, it was recently suggested that they be included in this series. No changes have been made in the original version, written by E. Halpern, to whom I am glad to express my hearty thanks. However, some comments have been added at the end, which, without aiming at completeness, point out further results and give more recent references.

A. Borel

TABLE OF CONTENTS

CHAPTER I

THE HOMOLOGICAL PROPERTIES OF H-SPACES

1. Algebraic preliminaries.

We introduce the following notations, terminology, and conventions which will be used throughout.

Z: the ring of integers,

Z_m: the ring of integers modulo m,

Z_o: the field of rational numbers,

K, K_p: a field of characteristic p.

Tors G: the torsion subgroup of an abelian group G. This consists of the elements of G which are of finite order.

$\mathrm{Tors}_p\, G$: the p-primary component. For p a prime this consists of the elements of G whose order are a power of p; for p = 0 $\mathrm{Tors}_o\, G$ consists of the zero element. If p is a prime and $\mathrm{Tors}_p\, G = 0$ we say that G has no p-torsion; it is also convenient to say that G has no o-torsion.

A: a ring (with unit 1).

M: an A-module (1 induces the identity on M). It is said to be graded if $M = \sum_i M^i$ (weak direct sum) where the M^i are A-modules. In all cases which we consider we shall assume $M^i = 0$ for $i < 0$, and that M^i is finitely generated over A. When A is a field the Poincaré polynomial (or Poincaré series if M is infinite dimensional) is

$$P(M, t) = \sum \dim M^i . t^i.$$

The degree $d^o x$ of a non zero element is the smallest integer j such that $x \in \sum_{i=0}^{j} M^i$. If $x \in M^i$, we say that it is homogeneous of degree i.

Similarly M is bigraded if $M = \sum M^{i,j}$. An element $x \in M^{i,j}$ is said to be bihomogeneous of bidegree (i,j).

If M is an A-algebra then it is a graded A-algebra if it is graded as an A-module and $M^i M^j \subset M^{i+j}$. A unit (if any) must then be in M^o. Similarly, if it is bigraded as an A-module and satisfies $M^{i,j}.M^{i',j'} \subset M^{i+i',j+j'}$ we say it is a bigraded A-algebra.

If M is a graded A-algebra then it is anti-commutative (also skew commutative) if for homogeneous elements x and y we have

$$xy = (-1)^{d^o x \cdot d^o y} yx.$$

If $A = K_2$ then this is simply commutativity. If $A = K_p$ with $p \neq 2$ then for an element x of odd degree we have $x^2 = 0$.

If M_1, M_2 are graded A-modules then their <u>tensor product</u> $M_1 \otimes M_2$ is the usual tensor product bigraded by $M_1^i \otimes M_2^j$. We also have a <u>total grading</u> of $M_1 \otimes M_2$ obtained by defining the homogeneous elements of total degree n to be the elements of $\sum'_{i+j=n} M_1^i \otimes M_2^j$.

If M_1, M_2 are graded A-algebras then $M_1 \otimes M_2$ (as graded A-modules) is a bigraded A-algebra under the multiplication defined by

$$(x_1 \otimes x_2)(y_1 \otimes y_2) = (-1)^{d^o x_2 \cdot d^o y_1} (x_1 y_1) \otimes (x_2 y_2).$$

One verifies readily that if M_1, M_2 are anti-commutative then $M_1 \otimes M_2$ is anti-commutative with respect to the total degree.

ΛP: the Grassmann algebra of a vector space P. If P is finite dimensional and has $(x_1, \ldots, x_m)$ as basis then we denote it by $\Lambda(x_1, \ldots, x_m)$. If we regard it as an anti-commutative graded algebra then we assume the x_i to be homogeneous; if $p \neq 2$ then $x_i x_j = -x_j x_i$ implies that $d^o x_i$ is odd for all i.

An A-algebra M has a <u>simple system</u> of generators $(x_1, x_2, \ldots)$ if it is the (weak) direct sum of the monogenic A-modules generated by 1 (if M has a unit 1) and the monomials $x_{i_1} x_{i_2} \ldots x_{i_k}$ with $i_1 < i_2 \ldots < i_k$. If $(x_1, \ldots, x_m)$ is a simple system of generators for M we write $M = \Delta(x_1, \ldots, x_m)$. We cite the following examples:

(a) $\Lambda(x_1, \ldots, x_m)$;

(b) $A[x]$ has $(x, x^2, \ldots, x^{2^k}, \ldots)$ as a simple system of generators;

(c) $A[x]/(x^s)$ has a simple system of generators if and only if s is of the form 2^k $(k > 0)$ in which case $(x, x^2, \ldots, x^{2^{k-1}})$ is a simple system.

<u>Throughout we shall consider only associative algebras</u>.

2. <u>Topological preliminaries</u>.

We shall consider only topological spaces X which are arcwise connected. Most of the time we shall consider X to be a finite polyhedron and all homology theories will coincide. For more general spaces we shall consider either the singular or Čech homology theories. It is to be understood <u>once and for all</u> that X satisfies one of the following conditions:

(1) X is a finite polyhedron,

(2) X is a space with finitely generated singular homology groups (and hence also has finitely generated singular cohomology groups),

(3) X is a compact space and has finitely generated Čech cohomology groups (and hence also has finitely generated Čech homology groups).

As usual we denote the ith dimensional homology and cohomology groups of X with coefficients in the ring A by $H_i(X,A)$ and $H^i(X,A)$ respectively, and we set

$$H_*(X,A) = \sum H_i(X,A),$$
$$H^*(X,A) = \sum H^i(X,A).$$

The latter is the cohomology ring of X (under cup products) and is known to be an associative ring with a unit element, and moreover is anti-commutative in case A is commutative.

If $A = K_p$ then the homology and cohomology groups are finite dimensional vector spaces and the Poincaré polynomial of X (Poincaré series in cases (2) and (3)) is defined by

$$P_p(X,t) = \sum \dim H^i(X,K_p)t^i.$$

We recall the Künneth rule. If A is a module over a principal ideal domain L then the following sequence is exact:

$$0 \to \sum_{i+j=n} H^i(X,A) \otimes H^j(Y,A) \xrightarrow{\lambda} H^n(X \times Y,A) \to \sum_{i+j=n+1} \mathrm{Tor}(H^i(X,A),H^j(Y,A)) \to 0.$$

The map λ is given by $\lambda = \pi_1^* \otimes \pi_2^*$ where π_1 and π_2 are the natural projections of $X \times Y$. Note that π_1^* and π_2^* are monomorphisms (f). If $A = K$, or if $A = Z$ and either factor is torsion free, then Tor vanishes so that λ identifies $H^*(X,A) \otimes H^*(Y,A)$ with $H^*(X \times Y,A)$. In that case the map induced by $x \to (x,y_0)$ is given by

$$x \otimes 1 + \sum x_i \otimes y_i \to x, \quad (d^\circ y_i > 0),$$

and similarly for $y \to (x_0,y)$.

We shall use without further comment the fact that the Künneth rule always holds in cases (1) and (3), and also in case (2) when $A = K$, or when $A = Z$ and $H^*(X,Z)$ or $H^*(Y,Z)$ is torsion free. In case (2) we also have the above exact sequence for homology if we replace n+1 by n-1 in the last sum. In these various cases we also have the universal coefficient theorem which asserts the following sequences are exact:

$$0 \to H^i(X,L) \otimes A \to H^i(X,A) \to \mathrm{Tor}(H^{i+1}(X,L),A) \to 0,$$
$$0 \to H_i(X,L) \otimes A \to H_i(X,A) \to \mathrm{Tor}(H_{i-1}(X,L),A) \to 0.$$

(f) We shall use the following terminology. A map $f:A \to B$ is injective if $a \neq a'$ implies $f(a) \neq f(a')$; it is surjective if $f(A) = B$; it is bijective if it is injective and surjective. An injective (resp. surjective, bijective) homomorphism is a monomorphism (resp. epimorphism, isomorphism).

3. The structure of Hopf algebras.

A Hopf algebra consists of an anti-commutative graded algebra H (graded by non-negative degrees) with a unit element 1 which spans H^0 and a homomorphism $h:H \to H \otimes H$ such that if x is a homogeneous element with $d^0x > 0$ then

$$h(x) = x \otimes 1 + 1 \otimes x + \sum u_i \otimes v_i$$

where u_i and v_i are homogeneous elements such that $d^0x = d^0u_i + d^0v_i$ and $d^0u_i > 0$, $d^0v_i > 0$, the summation finite.

We shall denote a Hopf algebra by H although (H,h) would be more precise. An equivalent definition is to require instead that h satisfy

$$h(x) = \rho(x) \otimes 1 + 1 \otimes \sigma(x) + \sum u_i \otimes v_i$$

where ρ and σ are automorphisms of H: the equivalence of the definitions follows at once from that observation that if h satisfies the second condition then $(\rho^{-1} \otimes \sigma^{-1})h$ satisfies the first condition.

A monogenic Hopf algebra is generated by 1 and a homogeneous element x with $d^0x > 0$. The height of x is the integer s satisfying $x^{s-1} \neq 0$, $x^s = 0$. If no such integer exists then we define the height $s = \infty$. The following theorem gives a complete description of monogenic Hopf algebras.

Theorem 3.1. Let H be a monogenic Hopf algebra over a field of characteristic p.

(a) If $p \neq 2$ and d^0x is odd then $H = \Lambda(x)$.

(b) If $p \neq 2$ and d^0x is even then $s = p^k$ or ∞.

(c) If $p = 2$ then $s = 2^k$ or ∞.

Proof. (a) This is immediate since $x^2 = 0$.

(b) Since d^0x is even x is in the center of H. Then

$$h(x^s) = (x \otimes 1 + 1 \otimes x)^s = \sum \binom{s}{i} x^i \otimes x^{s-i}.$$

From the definition of s we know x^i and x^{s-i} are not 0. If the assertion is not true we can write $s = p^k m$ where $(m,p) = 1$ and $m > 1$. It follows easily that $\binom{s}{p^k} \equiv m \pmod p$. Thus in the above sum for $h(x^s)$ there is a non-zero term $mx^{p^k} \otimes x^{s-p^k}$ so that $h(x^s) \neq 0$. But then x^s cannot be 0 which contradicts the defining property of s.

(c) Since $p = 2$, H is commutative and the same argument as in (b) applies.

A system of generators of type (M) is a sequence of homogeneous elements $(x_1, x_2, \ldots)$ with the following properties:

(1) $(x_1, x_2, \ldots)$ is a minimal system of generators,

(2) $d^o x_i \leq d^o x_j$ for $i \leq j$,

(3) if $P = P(x_1, \ldots, x_{k-1})$ is any polynomial which represents a homogeneous element of degree $d^o x_k$ then the height of x_k satisfies

$$s_k \leq \text{height of } (x_k + P).$$

One proves readily that every Hopf algebra has a system of type (M) (in fact the existence of the homomorphism h is not required).

If K_p is a field with the property that it contains a p-th root of each of its elements then it is called <u>perfect</u>. Note that if $p = 0$ K_p is perfect.

We shall refer to graded algebras such that H^i is finitely generated for all i as algebras of <u>finite type</u>.

The main result of the section is the following structure theorem for Hopf algebras:

<u>Theorem 3.2</u>. If H is a Hopf algebra of finite type over a perfect field then it is isomorphic (as an algebra) to the tensor product of monogenic Hopf algebras.

In view of (3.1) this is an immediate consequence of the following theorem.

<u>Theorem 3.3</u>. Let H be a Hopf algebra of finite type over a perfect field K_p and let (x_i) be a system of generators of type (M). We conclude:

(1) The monomials $x_1^{r_1} x_2^{r_2} \ldots x_m^{r_m}$, where $0 \leq r_i < s_i$ $(i = 1,2,\ldots,m)$ with $r_i = 0$ except for a finite number of indices, form a vector basis for H.

(2) If s_i is the height of x_i and

(a) $p = 2$ then $s_i = 2^{k_i}$ or $s_i = \infty$,

(b) $p \neq 2$, $d^o x_i$ odd, then $s_i = 2$,

(c) $p = 0$, $d^o x_i$ even, then $s_i = \infty$,

(d) $p \neq 0,2$, $d^o x_i$ even, then $s_i = p^{k_i}$ or $s_i = \infty$.

From (2) it is clear that each x_i generates a monogenic Hopf algebra H_i under $h_i(x) = x \otimes 1 + 1 \otimes x$. Then $\otimes_i H_i$ is a Hopf algebra under $\otimes_i h_i$, and applying (1) we obtain the preceding theorem. In general h and $\otimes_i h_i$ are unrelated; this is the meaning of the parenthetical remark in theorem (3.2).

Note that if $p = 0$, H is isomorphic to the tensor product of an exterior algebra generated by elements of odd degree and a ring of polynomials generated by elements of even degree. If in addition H is finite dimensional then H is an exterior algebra generated by elements of odd degree. This is the original Hopf theorem [4].

Although we are primarily interested in Hopf algebras of finite type the theorems are probably true more generally.

We begin the proof of theorem 3.3 with some preliminaries (the proof appears in [1]).

If $d^o x$ is odd and $r > 1$ then $x^n = 0$ when $p \neq 2$; hence $x^r \neq 0$ with $r > 1$ will mean that $p = 2$ or if $p \neq 2$ that $d^o x$ is even. In either case x is the center of H.

If (x_i) is a given system of generators of type (M) for H we let I_k stand for the ideal $(x_1,\ldots,x_k) \otimes H$ in $H \otimes H$. Then we can write

$$h(x_k) \equiv x_k \otimes 1 + 1 \otimes x_k \quad (\text{mod } I_{k-1}),$$
$$h(x_i) \equiv 1 \otimes x_i \quad (\text{mod } I_{k-1}) \quad \text{for } i \leq k-1 :$$

hence, h being a homomorphism, if $r > 1$

$$h(x_k^r) \equiv (x_k \otimes 1 + 1 \otimes x_k)^r \equiv x_k^r \otimes 1 + \sum_{0 \leq i < r} \binom{r}{i} x_k^i \otimes x_k^{r-i} \quad (\text{mod } I_{k-1}),$$
$$h(x_{k-1}^{r_{k-1}} x_{k-2}^{r_{k-2}} \ldots x_1^{r_1}) \equiv 1 \otimes x_{k-1}^{r_{k-1}} x_{k-2}^{r_{k-2}} \ldots x_1^{r_1} \quad (\text{mod } I_{k-1}),$$
$$h(x_k^r Q) \equiv x_k^r \otimes Q + \sum_{0 \leq i < r} \binom{r}{i} (x_k^i \otimes x_k^{r-i})(1 \otimes Q) \quad (\text{mod } I_{k-1}),$$

where $Q = Q\,(x_1,\ldots,x_{k-1})$ is a polynomial.

A monomial

$$a = x_k^{r_k} x_{k-1}^{r_{k-1}} \ldots x_1^{r_1};\ 0 < r_k < s_k,\ 0 \leq r_i < s_i\ (1 \leq i < k)$$

is called <u>normal</u>. By the <u>degree of a</u> we mean $\sum_{1 \leq i \leq k} r_i d^o x_i$. If a,b are normal monomials then $a \otimes b$ is said to be a normal monomial in $H \otimes H$. It is easy to show that H is generated by normal monomials. Thus to prove (1) of theorem 3.3 it remains to show that the normal monomials are linearly independent. We prove this by induction on the degree of a normal monomial.

For degree $n = 1$ this is trivial. Assume it is true for degree less than $n (n > 1)$. Note that this means that any two linear combinations of normal monomials $a_1, a_2, \ldots, a_j$ of degrees less than n are equal as elements of H if and only if they are formally identical; in particular, the normal monomials $a_i \otimes b_j$ such that the degrees of the a_i and the b_j are less than n are linearly independent in $H \otimes H$. Now let $P = P\,(x_k,\ldots,x_1)$ be any linear combination of normal monomials of degree n with non-zero coefficients and such that $P = 0$. We shall produce a contradiction.

(α) <u>We assert that P can be written</u>

$$P(x_k,\dots,x_1) = x_k^r + R(x_k,\dots,x_1)$$

where the exponent of x_k in the polynomial R is less than r.

<u>Proof</u>. Clearly we can write

$$P(x_k,\dots,x_1) = x_k^r\, Q\,(x_{k-1},\dots,x_1) + R\,(x_k,\dots,x_1)$$

(everything written in lexicographic order) with the specified condition on the exponent of x_k in R. Suppose Q has positive degree; then $r\, d^o x_k < n$. From the above formula for $h(x_k^r Q)$ we see that it contains a normal monomial of the form $x_k^r \otimes \lambda a$ where $\lambda \neq 0$ and $a(\neq 0)$ is the greatest normal monomial in Q. Observe that by the restriction on the exponent of x_k in R there are no normal monomials in h(R) which can cancel this, a priori formally, but then also in $H \otimes H$ by the remark preceeding (α). Since the degrees of x_k^r and a are less than n and $\lambda \neq 0$, it follows that $h(P) \neq 0$. But this contradicts the fact that P = 0. Therefore Q cannot have positive degree. By dividing out Q the assertion is proved.

(β) <u>If the characteristic p = 0 then r = 1</u>.

If $r \neq 1$ then from the above formula for $h(x^r)$ we see that it contains at least one term of the form

$$\binom{r}{i} x_k^i \otimes x_k^{r-i},\ 0 < i < r,$$

which cannot be cancelled by any linear combination of terms in h(R). Since p = 0 and the degrees of both factors are less than n, we know that this term is non-zero, and hence also that $h(P) \neq 0$. But this contradicts P = 0; hence r must be 1.

We are now in a position to prove (1) of the main theorem (3.3) when the characteristic p = 0. We have

$$P(x_k,\dots,x_1) = x_k + R(x_k,\dots,x_1) = 0.$$

But this contradicts the fact that (x_i) is a minimal system. Thus the normal monomials of degree n are linearly independent.

(γ) <u>If the characteristic $p \neq 0$ then $r > 1$ and is a power of p</u>.

This follows by an argument similar to (β) from the property of binomial coefficients recalled in the proof of theorem 3.1.

(δ) <u>Every normal monomial in P can be written in the form μz^r where $\mu \in K_p$ and z is a monomial</u> (not necessarily normal).

The proof is by induction on the decreasing lexicographic order in P. Note that the first normal monomial in P is x_k^r which is of this form. Assume that we can write P in the form

$$P = x_k^r + S + x_j^t\, U(x_{j-1},\dots,x_1) + V(x_j,\dots,x_1)$$

where by inductive assumption $S = \sum \mu_i z_i^r$ and the exponent of x_j in V is less than t. Since the field is perfect and r is a power of p we may write

$$S = \sum (\bar{\mu}_i z_i)^r = \bar{S}^r, \text{ where } \bar{\mu}_i^r = \mu_i,\ \bar{S} = \sum \bar{\mu}_i z_i:$$

hence we have

$$P = (x_k + \bar{S})^r + x_j^t\, U(x_{j-1},\dots,x_1) + V(x_j,\dots,x_1).$$

We may write

$$h(x_k + \bar{S}) = (x_k + \bar{S}) \otimes 1 + 1 \otimes (x_k + \bar{S}) + \sum c_i a_i \otimes b_i,$$

$a_i \otimes b_i$ being independent normal monomials with $d^o a_i > 0$, $d^o b_i > 0$: therefore

$$h((x_k + \bar{S})^r) = (x_k + \bar{S})^r \otimes 1 + 1 \otimes (x_k + \bar{S})^r + \sum \pm c_i^r\, a_i^r \otimes b_i^r,$$

and the non-zero terms of the last sum are (up to coefficients) independent normal monomials.

Assume first that $d^o U > 0$; then we consider the term $\lambda x_j^t b$, where b is the greatest normal monomial in U. As in (α) we see that $h(x_j^t U)$ contains $\lambda x_j^t \otimes b$ which cannot be cancelled in $h(x_j^t U + V)$. Hence, always using the remark preceding (α), there exists an i such that $x_j^t = \pm a_i^r$, $b = \pm b_i^r$, whence $x_j^t b = (c_i a_i b_i)^r$ with $c_i^r = \pm 1$.

Assume now that U is a constant λ. If t is a power of p then it is divisible by r since $d^o x_j \leq d^o x_k$ implies $t \geq r$. If t is not a power of p we see by the now familiar argument that $h(\lambda x_j^t)$ contains a term $\mu x_j^s \otimes x_j^{t-s}$, $(0 < s < t,\ \mu \neq 0)$, which cannot be cancelled by $h(\lambda x_j^t + V)$ and must therefore be equal to one term $c_i a_i^r \otimes b_i^r$. It follows readily that $x_j^t = (c_i a_i b_i)^r$. This completes the proof of (δ).

We now prove (1) of theorem 3.3 in case $p \neq 0$. By (δ) we can write P in the form

$$P = x_k^r + \sum \mu_i z_i^r,\ \mu_i \in K_p,$$

where z_i is a monomial which does not contain x_k. Thus

$$P = (x_k + \sum_i \bar{\mu}_i\, z_i)^r = (x_k + \bar{S})^r = (x_k + \bar{S}(x_1,\dots,x_{k-1}))^r.$$

Since P = 0 we have $x_k + \bar{S}\,(x_1,\dots,x_{k-1}) = 0$, and hence, we have height $(x_k + \bar{S}) \leq r <$ height x_k. But this contradicts the fact that (x_i) is a system of type (M). Thus the induction to degree n is complete and (1) is proved.

Part (2) is proved similarly to theorem 3.1 using the following lemma. The details are left to the reader.

Lemma. Let (x_i) be a system of generators of type (M). If x_k is in the center of H, s is not a power of p, and $x_k^{s-1} \neq 0$ then $x_k^s \neq 0$.

Corresponding to each x_i we can write a Poincaré polynomial series

$$P_p(H_i,t) = \begin{cases} 1 + t^{d^o x_i} + t^{2d^o x_i} + \ldots + t^{(s_i-1)d^o x_i} & \text{if } s_i < \infty, \\ (1-t^{d^o x_i})^{-1} & \text{if } s_i = \infty. \end{cases}$$

(In the latter case we mean of course the infinite series.) The dimension of H_i is given by $P_p(H_i,1)$; hence $s_i = \dim H_i$. Therefore if H has finite dimension we have $\dim H = s_1 \cdot s_2 \ldots s_m$.

Proposition 3.4. Every Hopf algebra of finite type over a perfect field K_2 has a simple system of generators.

Let (x_i) constitute a system of generators of type (M). Then by (3.3) the elements $x_i^{2^j}$, $1 \leq 2^j < s_i$, form a simple system of generators.

Proposition 3.5. If H is a finite dimensional Hopf algebra over a perfect field K_p, $p \neq 2$, then the following are equivalent:

(a) $H = \Lambda(x_1,\ldots,x_m)$ with $d^o x_i$ odd,

(b) $(x_1,\ldots, x_m)$ is a simple system for H,

(c) $\dim H = 2^m$.

Clearly (a) → (b) → (c). It remains to show (c) → (a). Let $(x_1,\ldots,x_m)$ be a system of type (M) for H and let $s_1,\ldots,s_m$ be the respective heights. Then by (c) $s_1\ldots s_m = 2^m$ so that $s_i = 2$ for all i. This proves (a).

Proposition 3.6. If H is a finite dimensional Hopf algebra over a perfect field K_p and the Poincaré polynomial has the form

$$P(H,t) = (1 + t^{k_1})(1 + t^{k_2}) \ldots (1 + t^{k_m}) \text{ with } k_i \text{ odd}$$

then

$$H = \Lambda(x_1,\ldots,x_m) \text{ with } d^o x_i \text{ odd.}$$

For $p \neq 2$ this reduces to 3.5. Let p = 2. Any simple system of generators of H consists of m elements of degrees $k_1,k_2,\ldots,k_m$ respectively. But in the simple system constructed in the proof of 3.4 there are odd degrees only if the x_i have odd degrees and height 2. This proves the proposition.

Clearly our definition of Hopf algebra (over K_p) may be extended by considering Z (or any ring) in place of K_p. Little is known of the structure of such Hopf algebras. Even without torsion H may be complicated. As an example we cite $H^*(\Omega_{n+1}, Z)$ where Ω_{n+1} is the loop space of an odd dimensional sphere S_{n+1}. It is known that $H^*(\Omega_{n+1}, Z)$ is a twisted polynomial ring; explicitly,

$$H^i(\Omega_{n+1}, Z) = \begin{cases} 0 & \text{if } i \neq kn \\ Z & \text{if } i = kn, \text{ generator } e_k. \end{cases}$$

with multiplication given by

$$e_j\, e_k = \binom{j+k}{j} e_{j+k}.$$

<u>Theorem 3.7</u>. If H is a Hopf algebra over Z of finite rank with no torsion then

$$H = \Lambda\,(x_1,\ldots,x_m),\ d^o x_i \text{ odd (all i)}.$$

<u>Proof</u>. Let D^i be the group of decomposable elements in H^i and D_p^i the space of decomposable elements in $H^i \otimes Z_p$. We can choose a basis $(y_{i,1},\ldots,y_{i,s_i}, x_{i,1},\ldots,x_{i,t_i})$ for H^i such that for suitable integers $m_{ij} \neq 0$ the elements $m_{ij}y_{ij}$ form a basis for D^i. We regard $H \subset H \otimes Z_o$, in which case D^i generates D_o^i. Since H has finite rank so has $H \otimes Z_o$, and we know from the (Hopf) structure theorem that

$$H \otimes Z_o = \Lambda\,(x_1',\ldots,x_m'),\ d^o x_i' \text{ odd},$$

where $x_i' = x_i \otimes 1$. Thus we can write the Poincaré polynomial

$$P(H \otimes Z_o,t) = \Pi(1 + t^{d^o x_i'}),\ d^o x_i' \text{ odd}.$$

For $p \neq 0$ we know $H \otimes Z_p \approx H/pH$ with D^i mapped onto D_p^i. Then $H \otimes Z_p$ is a Hopf algebra under the homomorphism induced by h. Clearly

$$\dim H^i \otimes Z_p = \dim H^i \otimes Z_o = \operatorname{rank} H^i,$$

and hence it follows that

$$P(H \otimes Z_p,t) = P(H \otimes Z_o,t).$$

Therefore by proposition 3.6 it follows that

$$H \otimes Z_p = \Lambda\,(x_{p_1},\ldots,x_{p_m}),\ d^o x_{p_i} = d^o x_i'.$$

This implies that $\dim D_p^i = \dim D_o^i = \operatorname{rank} D^i$, and hence we have $(m_{ij},p) = 1$ for any p. Thus $m_{ij} = \pm 1$ and D^i is a direct summand, and the theorem is proved.

If H is a Hopf algebra over Z then H/Tors H is a Hopf algebra under the homomorphism naturally induced by h. Hence we have the following corollary.

<u>Corollary 3.8</u>. If H is a Hopf algebra over Z which is finitely generated then

$$H/\text{Tors } H = \Lambda\,(x_1,\ldots,x_m),\ d^o x_i \text{ odd}.$$

4. Primitive elements; Samelson's theorem.

Consider the following example. Let $H = \Lambda(x_1, x_2)$ over K_2 with $d^o x_1 = 1$, $d^o x_2 = 2$, and h defined by

$$h(x_1) = x_1 \otimes 1 + 1 \otimes x_1,$$
$$h(x_2) = x_2 \otimes 1 + 1 \otimes x_2 + x_1 \otimes x_1.$$

Then H is isomorphic to $H_1 \otimes H_2$ where

$$H_1 = \Lambda(x_1) \text{ with } h_1(x_1) = x_1 \otimes 1 + 1 \otimes x_1,$$
$$H_2 = \Lambda(x_2) \text{ with } h_2(x_2) = x_2 \otimes 1 + 1 \otimes x_2.$$

However there is clearly no isomorphism of H on $H_1 \otimes H_2$ carrying h onto $h_1 \otimes h_2$. Thus an algebra can have essentially distinct homomorphisms h.

Let $x \in H$ be a homogeneous element with $d^o x > 0$. We say x is primitive if $h(x) = x \otimes 1 + 1 \otimes x$. One sees readily that the uniqueness of h is equivalent with the existence of a system of primitive generators of type (M). We shall now discuss a particular case where we can obtain such a system.

We say h is associative if the following diagram is commutative

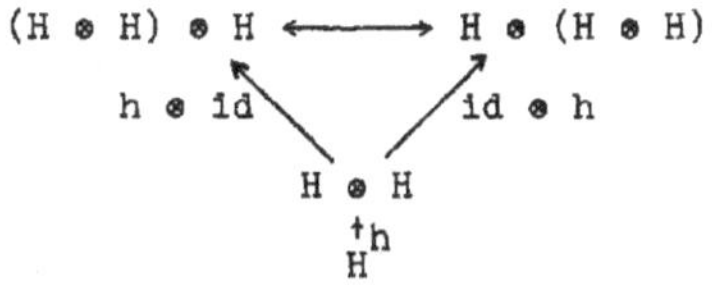

where id denotes the identity map and the horizontal map is the canonical isomorphism. If h is associative we say that H is an associative Hopf algebra. The following example shows that not every h is associative: $H = \Lambda(x_1, x_2, x_3)$ with $d^o x_1 = 1$, $d^o x_2 = 1$, $d^o x_3 = 3$, x_1 and x_2 primitive, and

$$h(x_3) = x_3 \otimes 1 + 1 \otimes x_3 + x_1 x_2 \otimes x_1.$$

Theorem 4.1. Let H be a Hopf algebra over K_p with associative h. If $H = \Lambda(x_1, \ldots, x_m)$ with $d^o x_i$ odd then there exist primitive elements $u_1, \ldots, u_m$ with $d^o u_i = d^o x_i$ such that $H = \Lambda(u_1, \ldots, u_m)$.

This is the algebraic formulation of Samelson's theorem; see (5.8) and (6.6). The proof given by H. Samelson [8] is restricted to characteristic 0. The proof we give follows a proof by J. Leray [6] which holds for any p.

Proof. Since x_1 is primitive we take $u_1 = x_1$. Assume there exist primitive elements $u_1, u_2, \ldots, u_k$ with $d^o u_i = d^o x_i$ and such that $H = \Lambda(u_1, \ldots, u_k, x_{k+1}, \ldots, x_m)$. We propose to find a primitive element u_{k+1} of the form

$$u_{k+1} = x_{k+1} + P(u_1, \ldots, u_k),$$

where $P(u_1, \ldots, u_k)$ is a polynomial of degree $d^o x_{k+1}$. This will prove the theorem.

We can write

$$h(x_{k+1}) = x_{k+1} \otimes 1 + 1 \otimes x_{k+1} + \sum c_{i_1 \ldots i_j \, ; \, i_{j+1} \ldots i_s} u_{i_1} \ldots u_{i_j} \otimes u_{i_{j+1}} \ldots u_{i_s}$$

where

(i) $i_1 < \cdots < i_j \; ; \; i_{j+1} < \cdots < i_s$,

(ii) $d^o u_{i_1} + \cdots + d^o u_{i_s} = d^o x_{k+1}$,

(iii) The summation is taken over all such $u_{i_1}, \ldots, u_{i_s}$ and all j such that $1 \leq j < s$.

Note that since $d^o x_{k+1}$ is odd, condition (ii) implies that $s \geq 3$.

The coefficients $c_{i_1 \ldots i_j \, ; \, i_{j+1} \ldots i_s}$ are here only defined for the special subscripts $i_1 < \cdots < i_j \; ; \; i_{j+1} < \cdots < i_s$. We extend the definition to all subscripts $i_1, \ldots, i_s$ by anti-symmetry in each batch of indices. We assert the following properties:

(α) $c_{i_1 \ldots i_j \, ; \, i_{j+1} \ldots i_s} = 0$ if the two batches have an integer in common.

(β) $c_{i_1 \ldots i_j \, ; \, i_{j+1} \ldots i_s} = c_{i_1 \ldots i_j i_{j+1} \, ; \, i_{j+2} \ldots i_s}$, $(s > j+1)$,

(γ) $c_{i_1 \ldots i_j \, ; \, i_{j+1} \ldots i_s} = -c_{i_1 \ldots i_{j-1} i_{j+1} \, ; \, i_j i_{j+2} \ldots i_s}$, $(s > j+1)$,

(δ) if $k_1 \ldots k_s$ is any permutation of $i_1 \ldots i_s$ then

$$\operatorname{sgn}(i_1 \ldots i_j \, ; \, i_{j+1} \ldots i_s) \, c_{i_1 \ldots i_j \, ; \, i_{j+1} \ldots i_s} = \operatorname{sgn}(k_1 \ldots k_m \, ; \, k_{m+1} \ldots k_s) \, c_{k_1 \ldots k_m ; k_{m+1} \ldots k_s}.$$

Assuming (δ) proved we can complete the proof of the theorem as follows. First note that if $v_1, \ldots, v_s$ are primitive elements then

$$h(v_1 \ldots v_s) = v_1 \ldots v_s \quad 1 + 1 \quad v_1 \ldots v_s + \sum_{1 \leq m < s} \operatorname{sgn}(k_1 \ldots k_m \, ; \, k_{m+1} \ldots k_s) \, v_{k_1} \ldots v_{k_m} \otimes v_{k_{m+1}} \ldots v_s.$$

Now let $c_{i_1 \ldots i_j \, ; \, i_{j+1} \ldots i_s} u_{i_1} \ldots u_{i_j} \otimes u_{i_{j+1}} \ldots u_s$ be a non zero fixed element in $h(x_{k+1})$. By (α) the indices are pairwise different ; therefore we may write

$$\operatorname{sgn}(i_1, \ldots, i_j \, ; \, i_{j+1}, \ldots, i_s) \, u_{i_1} \ldots u_{i_s} = u_{\mu_1} \ldots u_{\mu_s},$$

$\mu_1, \ldots, \mu_s$ being the permutation of the i_j with $\mu_1 < \cdots < \mu_s$. Using this and (δ) we see that the summation corresponding to

$h(\mathrm{sgn}(i_1\cdots i_j;i_{j+1}\cdots i_s)c_{i_1\cdots i_j;i_{j+1}\cdots i_s}u_{i_1}\cdots u_{i_s}$ is

$$\sum_{1\le m<s}\mathrm{sgn}(i_1\cdots i_j;i_{j+1}\cdots i_s)c_{i_1\cdots i_j;i_{j+1}\cdots i_s}\mathrm{sgn}(k_1\cdots k_m;k_{m+1}\cdots k_s)u_{k_1}\cdots u_{k_m}\otimes u_{k_{m+1}}\cdots u_{k_s}$$

$$= \sum_{1\le m<s} c_{k_1\cdots k_m;k_{m+1}\cdots k_s}\, u_{k_1}\cdots u_{k_m}\otimes u_{k_{m+1}}\cdots u_{k_s}$$

where the summation is extended over the permutations of $i_1,\ldots,i_s$ with $k_1<\cdots<k_m;k_{m+1}<\cdots<k_s$. From this it follows that h applied to

$$x_{k-1} - \mathrm{sgn}(i_1\cdots i_j;i_{j+1}\cdots i_s)c_{i_1\cdots i_j;i_{j+1}\cdots i_s}u_{i_1}\cdots u_{i_s}$$

contains strictly less elements of the form $a \otimes b$, $(d^0a > 0,\ d^0b > 0)$, than $h(x_{k+1})$. The construction of u_{k+1} then follows easily by an inductive argument.

It remains to prove $(\alpha,\beta,\gamma,\delta)$.

(α) Suppose $(i_1,\ldots,i_j)$ and $(i_{j+1},\ldots,i_s)$ have an integer in common. Using antisymmetry if necessary, we may assume that i_j is the common integer. Hence $h(x_{k+1})$ contains a term $c_{i_1\cdots i_j;i_{j+1}\cdots i_s}u_{i_1}\cdots u_{i_j}\otimes u_{i_{j+1}}\cdots u_{i_s}$ which contributes the following term to the image of $h \otimes \mathrm{id}$:

$$c_{i_1\cdots i_j;i_{j+1}\cdots i_s}(u_{i_1}\cdots u_{i_{j-1}}\otimes u_{i_j})\otimes u_{i_{j+1}}\cdots u_{i_s}$$

and, moreover, is the only one which does so. By the associativity of h,

$$c_{i_1\cdots i_j;i_{j+1}\cdots i_s}u_{i_1}\cdots u_{i_{j-1}}\otimes(u_{i_j}\otimes u_{i_{j+1}}\cdots u_{i_s})$$

is then in $\mathrm{id}\otimes h$. Thus $c_{i_1\cdots i_j;i_{j+1}\cdots i_s}\,u_{i_j}\otimes u_{i_{j+1}}\cdots u_{i_s}$ must be in the image of $h(P(u_1\cdots u_k))$. Since u_{i_j} appears on both sides of $\otimes$ we see that this impossible. Hence $c_{i_1\cdots i_j;i_{j+1}\cdots i_s} = 0$.

(β) As above it suffices to prove the assertion for the coefficients of the term in $h(x_{k+1})$. If $c_{i_1\cdots i_j;i_{j+1}\cdots i_s}u_{i_1}\cdots u_{i_j}\otimes u_{i_{j+1}}\cdots u_{i_s}$ is in $h(x_{k+1})$ then it contributes the following term to $\mathrm{id}\otimes h$:

$$c_{i_1\cdots i_j;i_{j+1}\cdots i_s}u_{i_1}\cdots u_{i_j}\otimes(u_{i_{j+1}}\otimes u_{i_{j+2}}\cdots u_{i_s}).$$

By associativity of h we can see that the following term is in $h\otimes\mathrm{id}$:

$$c_{i_1\cdots i_j;i_{j+1}\cdots i_s}(u_{i_1}\cdots u_{i_j}\otimes u_{i_{j+1}})\otimes u_{i_{j+2}}\cdots u_{i_s}.$$

But then it must be contributed by the following term which proves (β):

$$c_{i_1\cdots i_j i_{j+1};i_{j+2}\cdots i_s}u_{i_1}\cdots u_{i_{j+1}}\otimes u_{i_{j+2}}\cdots u_{i_s}.$$

(γ) This follows easily from (β).

(δ) This follows easily from (β) and (γ).

The same proof applies to the following theorem.

Theorem 4.2. Let H be a Hopf algebra over Z, without torsion, and with associative h. If $H = \Lambda(x_1,\ldots,x_m)$ with $d^o x_i$ odd then there exist primitive elements $u_1,\ldots,u_m$ with $d^o u_i = d^o x_i$ such that $H = \Lambda(u_1,\ldots,u_m)$.

Proposition 4.3. If H is a Hopf algebra over Z with associative h and finitely generated then $H / \text{Tors } H = \Lambda(u_1,\ldots,u_m)$, the u_i primitive with respect to the map induced by h.

5. The Pontrjagin product ([2],[3],[5],[7]).

Let H^* denote a Hopf algebra (of finite type) over a field K_p. Let H_i be the dual space to H^i and define $H_* = \sum H_i$. The duality between H^* and H_* is expressed as usual by

$$\langle x,u\rangle = u(x),\ x \in H^*,\ u \in H_*.$$

This induces a duality between $H^* \otimes H^*$ and $H_* \otimes H_*$ by

$$\langle a \otimes b, u \otimes v\rangle = \langle a,u\rangle \langle b,v\rangle.$$

Corresponding to H^* and H_* we have the homomorphisms

$$H^* \xrightarrow{h^*} H^* \otimes H^* \qquad\qquad H_* \otimes H_* \xrightarrow{h_*} H_*$$

where h^* is the Hopf homomorphism of H^* and h_* is the transpose of h^*. Explicitly h_* is defined by

$$\langle x, h_*(u \otimes v)\rangle = \langle h^*(x), u \otimes v\rangle.$$

(Note that $H^* \otimes H^*$ is the tensor product of graded algebras but $H_* \otimes H_*$ is only the tensor product of graded modules). The element $h_*(u \otimes v)$ is called the Pontrjagin product of u and v and will be denoted by $u \otimes v$. The following properties of $\vee$ are readily seen:

(a) There is a unit element in H_* which spans H_o,

(b) $d^o(u \vee v) = d^o u + d^o v$,

(c) it is distributive in each variable,

(d) if h^* is associative then $\vee$ is associative.

We prove (d) as follows. We have

$$\langle x,(u \vee v) \vee w\rangle = \langle (h^* \otimes \text{id})\cdot h^*(x),\ u \otimes v \otimes w\rangle,$$

$$\langle x, u \vee (v \vee w)\rangle = \langle (id \otimes h^*)\cdot h^*(x),\ u \otimes v \otimes w\rangle.$$

If

$$(h^* \otimes id)\cdot h^*(x) = \sum_i a_i \otimes (b_i \otimes c_i)$$

then

$$\langle x, (u \vee v) \vee w\rangle = \sum_i \langle a_i, u\rangle\ \langle b_i, v\rangle\ \langle c_i, w\rangle = \langle x, u \vee (v \vee w)\rangle$$

which proves (d).

Hereafter we assume h^* is associative. We shall prove some relations between H^* and H_*.

Proposition 5.1. An element $x \in H^*$ is primitive if and only if it is orthogonal to the decomposable elements of H_*.

Proof. Suppose x is primitive. We have

$$\langle x, u \vee v\rangle = \langle h^*(x),\ u \otimes v\rangle = \langle x \otimes 1 + 1 \otimes x,\ u \otimes v\rangle = 0$$

since $d^o u$, $d^o v > 0$. Conversely, suppose x is orthogonal to the decomposable elements. If x is not primitive then we can write

$$h^*(x) = x \otimes 1 + 1 \otimes x + \sum a_i \otimes b_i$$

there the a_i are linearly independent and similarly the b_i. Choose $u, v \in H_*$ such that

$$\langle a_1, u\rangle = 1 = \langle b_1, v\rangle\ ;\ \langle a_i, u\rangle = 0 = \langle b_i, v\rangle \text{ for } i \geq 2.$$

Then we have

$$\langle x, u \vee v\rangle = \langle h^*(x),\ u \otimes v\rangle = 1$$

which contradicts the fact that x is orthogonal to all decomposable elements. Therefore x is primitive.

We remark that the same argument is valid if we consider H^* a Hopf algebra over Z (or a principle ideal ring L) with no torsion.

Let θ be the automorphism of $H^* \otimes H^*$ defined by

$$\theta\ (a \otimes b) = (-1)^{d^o a \cdot d^o b}\ (b \otimes a),\quad a, b \text{ homogeneous.}$$

We say that h^* is symmetric if $\theta\, h^* = h^*$.

Proposition 5.2. H_* is anti-commutative if and only if h^* is symmetric.

For the proof of 5.2 see [2].

Lemma. If x is a primitive element in the center of H^* then for any $u \in H_*$ we have

$$\langle x^n, u^n\rangle = n!\,(\langle x, u\rangle)^n$$

This may be proved easily by induction.

Proposition 5.3. If $H^* = Z[x]$ with $d^o x$ even then H_* is the twisted polynomial ring in "one" variable u with $d^o u = d^o x$. More precisely; if $d^o x = n$ then

$$H_i = \begin{cases} 0 \text{ if } i \neq kn, \\ Z \text{ if } i = kn, \text{ generator } u_k, \end{cases}$$

and multiplication $u_i \cdot u_j = \binom{i+j}{i} u_{i+j}$. The converse is also true.

Proof. Define the u_j by $\langle x^i, u_j \rangle = \delta_{ij}$. By the lemma

$$\langle x^i, u_1{}^i \rangle = i!(\langle x, u_1 \rangle)^i = i! \;;$$

hence $u_i = u_1{}^i/i!$. Therefore

$$u_i \cdot u_j = (u_1{}^i/i!)(u_1{}^j/j!) - u_1{}^{i+j}/i!j! = \binom{i+j}{i} u_{i+j}.$$

The converse, as well as the following proposition, may be proved similarly.

Proposition 5.4. If H^* is a twisted polynomial ring in "one" variable x of even degree and is generated by primitive elements then $H_* = Z[u]$ with $d^o u = d^o x$.

Proposition 5.5. If $H^* = Z_p[x]$, $p \neq 0$, and x in the center, then H_* is commutative and $H_* = H_1 \otimes H_2 \otimes \cdots$ where H_i is generated by 1 and u_i with $u_i{}^p = 0$ and $d^o u_i = p^i d^o x$.

Proof. We choose the generators u_i as follows. Let u_1 be dual to x. By the lemma $\langle u_1{}^n, x^n \rangle = n!$ so that $\langle u_1{}^i, x^i \rangle \neq 0$ for $i < p$ and $\langle u_1{}^p, x^p \rangle = 0$. Thus $u_1{}^p = 0$ and $d^o u_1 = pd^o x$. Now choose u_2 so that $\langle u_2, x^p \rangle = 1$; then $\langle u_2{}^i, x^{pi} \rangle \neq 0$ for $i < p$ and $\langle u_2{}^p, x^{p^2} \rangle = 0$. Thus $u_2{}^p = 0$ and $d^o u_2 = p^2 d^o x$. Continuing we obtain infinitely many generators u_i with the asserted properties, as is easily seen. Since x is primitive it is orthogonal to the decomposable elements; hence $H_* = H_1 \otimes H_2 \otimes \cdots$ where H_i is generated by 1 and u_i. Since x is in the center commutativity of H_* is clear.

Proposition 5.7. $H^* = \Lambda(x_1, \ldots, x_m)$ with x_i primitive if and only if $H_* = \Lambda(u_1, \ldots, u_m)$ where $(u_1, \ldots, u_m)$ and $(x_1, \ldots, x_m)$ are dual bases and $d^o u_i = d^o x_i$.

Proof. Assume first that $H^* = \Lambda(u_1, \ldots, u_m)$. Let $(x_{i_1 \ldots i_k})$ be a dual basis to the basis $(u_{i_1} \vee \cdots \vee u_{i_k})$, $i_1 < \cdots < i_k$, $1 \leq k \leq m$, for H_* . By construction the elements x_i are orthogonal to the decomposable elements of H_* and hence are primitive. It follows that

$$\langle x_{i_1} x_{i_2}, u_{i_1} \vee u_{i_2} \rangle = 1 \;, \; (i_1 < i_2),$$
$$\langle x_{i_1} x_{i_2}, u_{j_1} \vee \cdots \vee u_{j_k} \rangle = 0 \quad \text{otherwise}.$$

Using induction on k one sees that the elements $x_{i_1} \cdots x_{i_k}$, $(i_1 < \cdots < i_k)$, $(1 \leq k \leq m)$, form a basis for H^* which is dual to $(u_{i_1} \vee \cdots \vee u_{i_k})$. Thus $(x_{i_1} \cdots x_{i_k})$ is a simple system of primitive generators for H^* .

The converse is proved analogously; see[5].

Combining (5.7) with Samelson's theorem (4.1) we obtain a second algebraic version of Samelson's theorem:

<u>Theorem 5.8</u>. If $H^* = \Lambda(x_1, \ldots x_m)$ with $d^o x_i$ odd then $H_* = \Lambda(u_1, \ldots, u_m)$ and $H_* \approx H^*$ (and hence H_* is anti-commutative.)

A similar proof gives the same result for a Hopf algebra H^* over Z without torsion. Combining this with (3.7) we get

<u>Proposition 5.9</u>. If H^* is a finitely generated Hopf algebra over Z then $H_* /$ Tors $H_* = \Lambda(u_1, \ldots, u_m)$, $d^o u_i$ odd.

6. <u>Homology of H-spaces</u>.

Let X be a topological space and $h: X \times X \to X$, $h(x,y) = x \cdot y$, a continuous map. Let x_o be a fixed point in X; then h is an <u>A-essential product</u> for X if $x \cdot x_o$ and $x_o \cdot x$ induce automorphisms ρ^* and σ^* of $H^*(X,A)$. Since X is assumed to be arcwise connected ρ^* and σ^* are clearly independent of the choice of x_o.

The standard examples of spaces with A-essential products are H-spaces. A topological space X is an <u>H-space</u> if it has a product $X \times X \to X$ with a homotopy unit; explicitly, there is an element $e \in X$ such that $e \cdot e = e$ and the maps $x \to x \cdot e$ and $x \to e \cdot x$ are homotopic to the identity map of X relative to e.

It is obvious then that an H-space has an A-essential product for any A, and moreover if we take $x_o = e$ then ρ^* and σ^* are the identity automorphism.

<u>Theorem 6.1</u>. If X has a K_p-essential product h then $H^*(X,K_p)$ is a Hopf algebra under the induced map h^*. Moreover there exists a system of generators of type (M) which have properties (1) and (2) stated in theorem (3.3).

<u>Proof</u>. Using the Künneth rule we may identify $H^*(X,K_p) \otimes H^*(X,K_p)$ and $H^*(X \times X,K_p)$ by the map λ (see section 2). One sees readily that if x is a homogeneous element of $H^*(X,K_p)$ then

$$h^*(x) = \rho^*(x) \otimes 1 + 1 \otimes \sigma^*(x) + \sum u_i \otimes v_i$$

where u_i and v_i have positive degrees less than $d^o x$, and $d^o u_i + d^o v_i = d^o x$. Thus h^* is a Hopf homomorphism.

Using the universal coefficient theorem we may identify $H^*(X,K_p)$with $H^*(X,Z_p) \otimes K_p$. It then follows that $H^*(X,Z_p)$ is a Hopf algebra under the restriction of h^*. But Z_p is a perfect field so we may apply theorem 3.3 to $H(X,Z_p)$. A system of generators of type (M) for $H^*(X,K_p)$ will therefore have the desired properties.

As immediate corollaries we have:

Proposition 6.2. If $p = 2$ then the Hopf algebra $H^*(X, K_2)$ (under h^*) has a simple system of generators.

Proposition 6.3. If $p \neq 2$ and X is a finite polyhedron with a K_p-essential product then $H^*(X,K_p)$ has a simple system of generators if and only if

$$H^*(X,K_p) = \Lambda(x_1,..,x_m),\ d^o x_i \text{ odd (all } i).$$

Making use of some previous results (section 3) we can consider the case where X has a Z-essential product.

Proposition 6.4. If X is a finite polyhedron with no p-torsion and X has a Z-essential product then

$$H^*(X,K_p) = \Lambda(x_1,\dots,x_m),\ d^o x_i \text{ odd (all } i).$$

Proof. The Z-essential product h is clearly also a Z_o-essential product. Since X is finite dimensional it follows from the Hopf theorem that

$$H^*(X,Z_o) = \Lambda(x_1,\dots,x_m),\ d^o x_i \text{ odd (all } i).$$

Moreover since X has no p-torsion h will also be a Z_p-essential product and $P_o(x,t) = P_p(x,t)$. Applying (3.6) we get that

$$H^*(X,Z_p) = \Lambda(\bar{x}_1,\dots,\bar{x}_m),\ d^o \bar{x}_i \text{ odd (all } i).$$

Identifying $H^*(X,K_p)$ with $H^*(X,Z_p) \otimes K_p$ the proposition then follows.

In view of (3.7) and(3.8) we have the following:

Proposition 6.5. Let X have a Z-essential product h.

(a) If $H^*(X,Z)$ is free and finitely generated then

$$H^*(X,Z) = \Lambda(x_1,\dots,x_m),\ d^o x_i \text{ odd (all } i).$$

(b) If $H^*(X,Z)$ is finitely generated then

$$H^*(X,Z) \,/\, \mathrm{Tors}\, H^*(X,Z) = \Lambda(\bar{x}_1,\dots,\bar{x}_m), d^o \bar{x}_i \text{ odd (all } i),$$

where the quotient is regarded as a Hopf algebra under the homomorphism naturally induced by h^*.

Let X be an H-space with an associative product h. Then one sees easily that h^* is associative as defined in section 4. Using the Künneth rule we identify $H_*(X \times X,K_p)$ with $H_*(X,K_p) \otimes H_*(X,K_p)$; hence h induces a map

$$h_*: H_*(X,K_p) \otimes H_*(X,K_p) \to H_*(X, K_p).$$

If $u, v \in H_*(X,K_p)$ then $h_*(u \otimes v)$ is called the (topological) Pontrjagin product of u and v. Since h^* is associative and h_* and h^* are mutually transposes it follows that

$h_*(u \otimes v)$ is precisely the (algebraic) Pontrjagin product as defined in section 5. The same is true of the topological and algebraic Pontrjagin products naturally defined in $H_*(X,Z)$ / Tors $H_*(X,Z)$.

Combining (6 4) and (6.5) with (5.7, 5.8, 5.9) we get the following theorem where (a) with p = 0 is the topological formulation of Samelson's theorem.

Theorem 6.6. Let X be a finite polyhedron which is an H-space with an associative product h.

(a) If X has no p-torsion then

$$H^*(X,K_p) = \Lambda(x_1,\dots,x_m),\ d^o x_i \text{ odd},$$

$$H_*(X,K_p) = \Lambda(u_1,\dots,u_m),$$

where $H_*(X,K_p)$ is the Pontrjagin ring. The elements x_i are primitive under h^*, and the u_i are a dual basis to the x_i with $d^o u_i = d^o x_i$.

(b) If X has no torsion then in (a) we may replace K_p by Z.

(c) If X has no torsion then (a) is still valid if we replace $H^*(X,K_p)$ and $H_*(X, K_p)$ by the quotients $H^*(X,Z)$ / Tors $H^*(X,Z)$ and $H_*(X,Z)$ / Tors $H_*(X,Z)$ respectively. The x_i are then primitive relative to the homomorphism naturally induced by h^*.

7. Spaces on which an H-space operates.

We assume throughout that our H-spaces have an associative product.

An H-space X operates (on the right) on a topological space E if there is a continuous map $\phi : E \times X \to E$ such that(writing $(a,x) = ax$)

(i) $a(x\cdot y) = (ax)\cdot y$,

(ii) $ae = a$ up to a given homotopy,

where e is the unit of X.

We consider homology (and cohomology) with coefficients in a field K (which we shall omit writing). Then the Künneth rule gives an isomorphism

$$H_*(E) \otimes H_*(X) \to H_*(E \times X)$$

which together with the induced map ϕ_* gives a map

$$\phi_* : H_*(E) \otimes H_*(X) \to H_*(E).$$

We call this pairing the generalized Pontrjagin product and write $\phi_*(a \otimes x) = a \vee x$.

We define a dual pairing (thus a generalized "Pontrjagin cap" product) as follows. In a similar fashion to above we obtain a map

$$\phi^*: H^*(E) \to H^*(E) \otimes H^*(X).$$

For $a \in H^*(E)$, $u \in H_*(X)$ we define $a \vee u \in H^*(E)$ by

$$\langle a \vee u, v\rangle = \langle a, v \vee u\rangle, \; v \in H_*(E),$$

where the second $\vee$ denotes the generalized Pontrjagin product. In detail; if

$$\phi^*(a) = \sum a_i \otimes y_i \in H^*(E) \otimes H^*(X)$$

then we set

$$a \vee u = \sum a_i \langle y_i, u\rangle .$$

This clearly defines a pairing $H^k(E)$, $H_s(X) \to H^{k-s}(E)$ with the properties:

(a) $a \vee 1 = a$, where 1 is the unit of $H_*(X)$,

(b) $a \vee (u \vee v) = (a \vee u) \vee v$,

(c) if X operates on E and E', and $f : E \to E'$ commutes with the operations of X then

$$f^*(a' \vee u) = (f^*a') \vee u, \; a' \in H^*(E'), \; u \in H_*(X)$$

(d) if $u \in H_*(X)$ is orthogonal to all decomposable elements of $H^*(X)$ then

$$(a \cdot b) \vee u = a\cdot(b \vee u) + (-1)^{d^o u \cdot d^o b} \cdot (a \vee u)\cdot b$$

where $\cdot$ means the ordinary cup product.

Trom the existence of a (homotopy) unit and the remarks at the end of section 2 it follows that

$$\phi^*(a) = a \otimes 1 + \sum a_i \otimes y_i, \; 0 < d^o a_i < d\, a ,$$

whence property (a). Property (b) follows from the condition $a(x \cdot y) = (a \cdot x)y$. Property (c) is the dual of

$$f_*(a \vee u) = (f_* a) \vee u, \; a \in H_*(E), \; u \in H_*(X)$$

and is obvious.

To prove (d) choose an additive basis $(\bar u, \bar u_1, \dots, \bar u_m, \bar v_1, \dots, \bar v_n)$ such that

$$\langle \bar u, u\rangle = 1, \; \langle \bar u_i, u\rangle = 0, \; \langle \bar v_i, u\rangle = 0$$

and such that the $\bar v_i$ span the decomposable elements. Relative to this basis we can write

$$\phi^*(a) = a \otimes 1 + a_o \otimes \bar u + \sum a_i \otimes \bar u_i + \sum \bar a_i \otimes \bar v_i,$$

$$\phi^*(b) = b \otimes 1 + b_o \otimes \bar u + \sum b_i \otimes \bar u_i + \sum \bar b_i \otimes \bar v_i.$$

Therefore

$$\phi^*(ab) = ab \otimes 1 + (ab_o + (-1)^{d^o u \cdot d^o b} a_o b) \otimes \bar u + R,$$

where each term in R has a right factor which is decomposable or involves some of the $\bar u_i$. Applying the definitions of $a \vee u$, $b \vee u$, $(ab) \vee u$, (d) follows.

As an example let $E = X$ and let ϕ be the H-space product $h : X \times X \to X$. If $h^*(x) = \sum x_i \otimes y_i$ then $x \vee u = \sum x_i \langle y_i, u \rangle$. Suppose now $H^*(X) = \Delta(x_1, \ldots, x_m)$ with the x_i primitive. Then we know by a previous theorem that $H_*(X) = \Lambda(u_1, \ldots, u_m)$ with the (u_i) dual to (x_i). Clearly the u_i are orthogonal to the decomposable elements of $H^*(X)$. Since the monomials $x_{i_1} \cdots x_{i_k}$, $i_1 < \cdots < i_k$ generate H^* the pairing is completely determined by the $x_{i_1} \cdots x_{i_k} \vee u_i$. These products are given by

$$x_{i_1} \cdots x_{i_k} \vee u_i = \begin{cases} \operatorname{sgn}(i_1, \ldots, i_k i_s)\, x_{i_1} \cdots \hat{x}_{i_s} \cdots x_{i_k} & \text{if } i = i_s, \\ 0 \text{ if } i = i_1, \ldots, i_k, \end{cases}$$

where ^ means as usual that the variable is to be omitted. To see this, note that since the elements x_i are primitive we can write

$$h^*(x_{i_1} \cdots x_{i_k}) = \sum \operatorname{sgn}(\mu_{i_1} \cdots \mu_{i_s}; \mu_{i_{s+1}} \cdots \mu_k)\, x_{\mu_1} \cdots x_{\mu_s} \otimes x_{\mu_{s+1}} \cdots x_{\mu_k}$$

where the summation is taken over the permutations described earlier with the additional terms $1\, x_{i_1} \cdots x_{i_k}$ and $x_{i_1} \cdots x_{i_k}\, 1$ (for $s = o, k$). Then

$$x_{i_1} \cdots x_{i_k} \vee u_i = \sum \operatorname{sgn}(\mu_{i_1} \cdots \mu_{i_s}; \mu_{i_{s+1}} \cdots \mu_{i_k}) x_{\mu_1} \cdots x_{\mu_s} \langle x_{\mu_{s+1}} \cdots x_{\mu_k}, u_i \rangle .$$

If $i \neq i_1, \ldots, i_k$ then $\langle x_{\mu_{s+1}} \cdots x_{\mu_k}, u_i \rangle = 0$. If i is among $i_1, \ldots, i_k$, say $i = i_j$, then all $\langle x_{\mu_{s+1}} \cdots x_{\mu_k}, u_i \rangle = 0$ except $\langle x_{\mu_j}, u_i \rangle = 1$. Therefore the sum reduces to

$$x_{i_1} \cdots x_{i_k} \vee u_i = \operatorname{sgn}(\mu_1, \ldots, \mu_k, \mu_j)\, x_{\mu_1} \cdots \hat{x}_{\mu_j} \cdots x_{\mu_k} .$$

<u>Lemma 7.1.</u> Let $H^*(X, K_p) = \Delta(x_1, \ldots, x_m)$ with x_i primitive . If P is the subspace additively generated by the x_i then any base of P is a simple system of generators.

If $p \neq 2$ then we know by a previous theorem that $H^*(X, K_p) = \Lambda(x_1, \ldots, x_m)$, and the lemma is trivial. If $p = 2$ the lemma may be proved as follows. Let $y_1, \ldots, y_m$ be any base of P, then we can write each x_i as a linear combination of y_i. This means that each generating monomial $x_{i_1} \cdots x_{i_k}$ $(i_1 < \cdots < i_k)$ corresponds to a polynomial in the monomials $y_{i_1}^{r_1} \cdots y_{i_k}^{r_k}$. It therefore suffices to prove that any such monomial is a linear combination of generating monomials of the form $y_{j_1} \cdots y_{j_k}$ $(j_1 < \cdots < j_k)$. For $k = 1$ this is trivial. We then proceed by induction on the number of factors $r_1 + \cdots + r_k$. Since $p = 2$ it follows that the square of the primitive element y_i is also primitive and hence is in P. Therefore, if y_i appears as a factor in $y_{i_1}^{r_1} \cdots y_{i_k}^{r_k}$ with exponent greater than 1 we can replace y_i^2 by a linear combination of the y_j and so obtain a smaller number of factors in each term which arises. Then we can apply the inductive hypothesis and the lemma follows.

Theorem 7.2. Let X,E be finite polyhedra with X an H-space which operates on E. Regard X as operating on itself (on the right), and let $f:X \to E$ be a map which commutes with the operations of X on E and itself. If $H^*(X,K) = \Delta(x_1,\dots,x_m)$ with x_i primitive then the image of $H^*(E,K)$ under f^* is generated by primitive elements.

Proof. We have previously proved that $H_*(X,K) = \Lambda(u_1,\dots,u_m)$ where the u_i are dual to the x_i. Since f commutes with the operations we have

$$f^*(a \vee u_i) = (f^*a) \vee u_i,\ a \in H^*(E,K).$$

This shows that the image of f^* is stable under the $\vee$-product with any u_i. The theorem will then follow if we show that any subalgebra $A \subset H^*(X)$ which is stable under $\vee$-product with all u_i is generated by primitive elements.

Let P denote the subspace additively generated by the x_i; if we change the basis so that the first k basis elements additively generate $A \cap P$ then lemma 7.1 guarantees that the new basis is still a simple system. Hence we may assume that $x_1,\dots,x_k$ generate $A \cap P$. Let $a \in A$, then we can write

$$a = c_{i_1 \cdots i_s} x_{i_1} \cdots x_{i_s} + \text{other terms},$$

where $x_{i_1} \cdots x_{i_s}$ is the longest monomial appearing in a. By the above example we have that

$$x_{i_1} \cdots x_{i_s} \vee (u_{i_s} \vee \cdots \vee u_{i_2}) = x_{i_1}.$$

Since A is stable we have that $x_{i_1} \in A \cap P$. Similarly $x_{i_2},\dots,x_{i_s} \in P$, and hence the longest monomial of a is generated by basis elements of $A \cap P$. By induction we may show that every monomial in a is generated by basis elements of $A \cap P$. It follows that A is generated by the primitive basis elements $x_1,\dots,x_k$, and the theorem is proved.

Theorem 7.3. Let X,E be finite polyhedra and X an H-space which operates on E. Regard X as operating on itself (on the right), and let $f:X \to E$ be a map which commutes with the operations of X on E and itself. If $p \neq 2$ and

$$H^*(X,K_p) = \Lambda(x_1,\dots,x_m),\ d^0 x_i \text{ odd},$$

then $H^*(E,K_p) \approx A \otimes B$, where A and B are subalgebras of $H^*(E,K_p)$ such that f^* annihilates A and is injective on B.

Proof. By Samelson's theorem we may assume the elements x_i to be primitive.

Then $H_*(X,K_p) = \Lambda(u_1,\ldots,u_m)$ with (u_i) dual to (x_i). By the preceding theorem we have that $f^*H^*(E,K_p)$ is generated by primitive elements which we may assume to be $x_1,\ldots,x_s$. Thus $f^*H^*(E,K_p) = \Lambda(x_1,\ldots,x_s)$.

(a) We assert there exist elements $y_i \in H^*(E,K_p)$ with the properties:

$$f^*(y_i) = x_i \ , \ 1 \le i \le s,$$
$$y_i \vee u_j = \delta_{ij}, \ 1 \le i,\ j \le s.$$

Since f commutes with the operations of X there exist elements $\bar{y}_i \in H^*(E,K_p)$ such that

$$f^*(\bar{y}_i) = x_i \ , \ 1 \le i \le s,$$
$$\bar{y}_i \vee u_j = \delta_{ij}, \ 1 \le i \le j \le m.$$

Such $\bar{y}_i$ can always be constructed by property (c) of $\vee$. We now construct the elements y_i from the $\bar{y}_i$ by induction. Set $y_1 = \bar{y}_1$. Assume $y_1,\ldots,y_k$ chosen with the required properties; then we construct y_{k+1} also satisfying these properties as follows. Setting

$$y_{k+1,k} = \bar{y}_{k+1} - (\bar{y}_{k+1} \vee u_1 \vee\cdots\vee u_k)\, y_k\cdots y_1,$$

we get

$$f^*(y_{k+1,k}) = f^*(\bar{y}_{k+1}) - f^*(\bar{y}_{k+1}) \vee (u_1 \vee\cdots\vee u_k)\, f^*(y_k)\cdots f^*(y_1),$$
$$= x_{k+1} - x_{k+1} \vee (u_1 \vee\cdots\vee u_k)\, x_k\cdots x_1 = x_{k+1}.$$

We also have

$$y_{k+1,k} \vee u_j = \bar{y}_{k+1} \vee u_j - \bar{y}_{k+1} \vee (u_1 \vee\cdots\vee u_k)\, y_k\cdots y_1 \vee u_j,$$
$$= \delta_{k+1,j} \quad \text{for } k+1 \le j \le s.$$

Then

$$y_{k+1,k} \vee(u_1 \vee\cdots\vee u_k) = \bar{y}_{k+1} \vee(u_1 \vee\cdots\vee u_k) - [\bar{y}_{k+1} \vee(u_1 \vee\cdots\vee u_k)y_k\cdots y_1]\vee(u_1\vee\cdots\vee u_k)$$
$$= \bar{y}_{k+1} \vee(u_1 \vee\cdots\vee u_k) - \bar{y}_{k+1} \vee(u_1 \vee\cdots\vee u_k) = 0.$$

This implies that $y_{k+1,k}$ is annihilated by $\Lambda^k(u_1,\ldots,u_k)$.

Now assume that we have constructed an element $y_{k+1,t}$ with $t \le s$ with the following properties:

$$f^*(y_{k+1,t}) = x_{k+1},$$
$$y_{k+1,t} \vee u_j = \delta_{k+1,j} \quad \text{for } k+1 \le j \le s,$$
$$j_{k+1,t} \ \text{ is annihilated by } \ \Lambda^t(u_1,\ldots,u_k).$$

We then construct an element $y_{k+1,t-1}$ satisfying the same properties by setting

$$y_{k+1,t-1} = y_{k+1,t} - \sum_{\mu_1<\cdots<\mu_t\le k} y_{k+1,t} \vee (u_{\mu_1}\vee\cdots\vee u_{\mu_t})y_{\mu_t}\cdots y_{\mu_1}.$$

Thus by induction we see that there is an element $y_{k+1,1}$ having the properties:

$$f^*(y_{k+1,1}) = x_{k+1},$$

$$y_{k+1,1} \vee u_j = \delta_{k+1,j} \quad \text{for } 1 \leq j \leq s,$$

$y_{k+1,1}$ is annihilated by $\Lambda'(u_1,\ldots,u_k)$, or equivalently by each u_i. We therefore define $y_{k+1} = y_{k+1,1}$, and (α) is proved.

Now let A to be the subalgebra annihilated by $\Lambda(u_1,\ldots,u_s)$, and let B denote the subalgebra generated by $y_1,\ldots,y_s$.

(β) f^* is injective on B:

The degree of y_i is odd since x_i has odd degree. Since $p \neq 2$ it follows that $B = \Lambda(y_1,\ldots,y_s)$. Since $f^*(y_i) = x_i$, f^* is clearly injective.

(γ) We assert $A \cdot B = H^*(E,K_p)$:

Let x be an element annihilated by $\Lambda'(u_1,\ldots,u_s)$, then $x \in A$. Assume that if x is annihilated by $\Lambda^j(u_1,\ldots,u_s)$, $j > 1$, then $x \in A \cdot B$. Suppose y is annihilated by $\Lambda^{j+1}(u_1,\ldots,u_s)$, and consider

$$y' = y - \sum_{i_1<\cdots<i_j\leq s} (y \vee (u_{i_1} \vee\cdots\vee u_{i_j}))\, y_{i_j}\cdots y_{i_1}.$$

A straightforward computation shows that y' is annihilated by $\Lambda^j(u_1,\ldots,u_s)$. Similarly $y \vee (u_{i_1} \vee\cdots\vee u_{i_j})$ is annihilated by $\Lambda'(u_1,\ldots,u_s)$ so that $y \vee (u_{i_1} \vee\cdots\vee u_{i_j}) \in A$. Clearly then the sum is in $A \cdot B$. By the inductive assumption $y' \in A \cdot B$, and hence so is y.

(δ) A and B are linearly disjoint:

Given an element

$$h = \sum_{i_1<\cdots<i_j} a_{i_1\cdots i_j}\, y_{i_1}\cdots y_{i_j} = 0$$

with $a_{i_1\cdots i_j} \in A$, we have to show that $a_{i_1\cdots i_j} = 0$. Let $a_{1\ldots n}y_1\cdots y_n$ be the longest monomial in h. Then

$$h \vee (u_n \vee\cdots\vee u_1) = a_{1,\ldots,n}y_1\cdots y_n \vee (u_n \vee\cdots\vee u_1) = a_{1,\ldots,n}.$$

Since h is 0, we have $a_{1,\ldots,n} = 0$. By induction on the length of monomials we have that all $a_{i_1\cdots i_j} = 0$.

Combining (γ) and (δ) yields the desired result.

We mention two applications:

(i) Let X be a compact connected Lie group G, let $E = G/U$ be the left coset space of G modulo a closed subgroup U, and let $f: G \to G/U$ be the canonical projection. Then theorem 7.3 generalizes a theorem of Leray[7].

(ii) Let X be a compact connected Lie group G. Let E be a compact connected Lie group $\tilde{G}$ having a closed subgroup isomorphic to G. Let $f:G \to \tilde{G}$ be the injection map. Then theorem 7.3 generalizes a theorem of Samelson [8].

BIBLIOGRAPHY

[1] A. Borel, Sur la cohomologie des espaces fibrés principaux et des espaces homogènes de groupes de Lie compacts, Ann. of Math. 57 (1953),115-2o7.

[2] A. Borel, Sur l'homologie et la cohomologie des groupes de Lie compacts connexes, Amer. J. Math. 76 (1954), 273-342.

[3] R. Bott and H. Samelson, On the Pontrjagin product in spaces of paths, Comm.Math. Helv. 27 (1953), 32o-337.

[4] H. Hopf, Über die Topologie der Gruppenmannigfaltigkeiten und ihrer Verallgemeinerungen, Ann. of Math. 42 (1941), 22-52.

[5] T. Kudo, Homological structure of fibre bundles, Jour. Inst. of Polyt. Osaka City Univ. 2 (1952), 1o1-14o.

[6] J. Leray, Sur la forme topologique des espaces et sur les points fixes des représentations, J. Math. Pures Applic. 54 (1945), 95-167.

[7] J. Leray, Espaces où opère un groupe de Lie compact connexe, Comptes rendus de l'Académie des Science, 228 (1949), 1545-1547;
- - -,Applications continues commutant avec les éléments d'un groupe de Lie, ibid., 1784-1786.

[8] H. Samelson, Beiträge zur Topologie der Gruppen-Mannigfaltigkeiten, Ann.of Math.42 (1941), 1o91-1137.

CHAPTER II

SPECTRAL SEQUENCE OF A FIBRE BUNDLE

In this chapter we shall define spectral sequences and give the main properties of the spectral sequence of a fibre bundle. For the proofs of these statements we refer the reader to the following literature:

(i) the Čech approach [1], [5], [9],

(ii) the singular approach [11], [12],

(iii) in the case where the base space is a locally finite polyhedron [6], [7], [10].

Finally we shall make some simple applications.

8. Differential and filtered modules

Differential module: an A-module M together with an endomorphism $d:M \to M$ such that $dd = 0$. Relative to the differential d we define the derived module $H(M) = d^{-1}(0)/dM$. If M is graded then we require that d be homogeneous of degree r, meaning $dM_i \subset M_{i+r}$ where r is an integer independent of i. It follows that H(M) is then graded in a natural way. Similarly if M is bigraded we require that d be bihomogeneous; H(M) is then naturally bigraded.

Differential algebra: a differential module M with an automorphism w satisfying

$$dw + wd = 0$$

$$d(x\cdot y) = dx\cdot y + w(x)\cdot dy.$$

If M is graded (bigraded) then H(M) is a graded (bigraded) algebra. The standard cases are d of degree -1 (homology), or d of degree +1 (cohomology), and $w(x) = (-1)^{d^o x}x$.

Filtered modules: an increasing (decreasing) filtration of an A-module M is given by an increasing (decreasing) sequence of submodules M_i such that $M = \cup M_i$. For homology (cohomology) we usually consider the increasing (decreasing) case.

Corresponding to a decreasing filtration of M we define a filtration function f on M by setting f(x) equal to the maximum integer i (possibly ∞) such that $x \in M_i$. Clearly $f(0) = \infty$, $f(x+y) \geq \min(f(x),f(y))$, and $f(\alpha x) \geq f(x)$. Conversely given an integer valued function f(x) on M which satisfies these properties we can define a decreasing filtration for M by setting

$$M_i = \{x \mid f(x) \geq i\}.$$

Similarly an increasing filtration can be characterized by a filtration function.

With each filtered module M is associated a graded module:

$GrM = \sum M_i/M_{i+1}$, (decreasing case),

$GrM = \sum M_i/M_{i-1}$, (increasing case).

A decreasing filtration is <u>limited below</u> if $M_i = M$ for i small enough and <u>limited above</u> if $M_i = 0$ for i big enough. An increasing filtration is <u>limited below</u> if $M_i = 0$ for i small enough and <u>limited above</u> if $M_i = M$ for i big enough.

In what follows we shall only consider the decreasing case, an analogous discussion holds in the increasing case.

A <u>homomorphism</u> of filtered modules is a homomorphism $f : M \to M'$ such that $f(M_i) \subset M'_i$. Given such a homomorphism there is induced a homomorphism $f_w: GrM \to GrM'$.

<u>Lemma 8.1</u>. (a) If f_w is injective and if $\cap M_i = 0$ then f is also injective.

(b) If f_w is surjective and the filtration of M' is limited above then f is surjective.

<u>Proof.</u> (a) If $x \in M$, $x \neq 0$, there exists an integer j such that $x \in M_j$ but $x \notin M_{j+1}$. Then the projection $\bar{x} \in M_j/M_{j+1}$ of x is not zero. Therefore $f_w(\bar{x}) \neq 0$ which in turn implies $f(x) \neq 0$. Thus f is injective.

(b) Let $x \in M'$. Since the filtration of M' is limited above, for i big enough $M'_i \subset f(M)$. Suppose j is such that $M'_{j+1} \subset f(M)$; let $x' \in M'_j$ and let $\bar{x}'$ be its projection in M'_j/M'_{j+1}. Since f_w is surjective there is an $x \in M_j$ such that if $\bar{x}$ is its projection in M_j/M_{j+1} we have $f_w(\bar{x}) = \bar{x}'$. Then $x'-f(x) \in M'_{j+1}$, and hence there is a $y \in M$ with $x'-f(x) = f(y)$. Thus $x' \in f(M)$ and (b) follows by induction.

<u>Filtered algebra</u>: if M is an A-algebra which is filtered as an A-module we further require that $M_i \cdot M_j \subset M_{i+j}$. For a filtered algebra, GrM is an algebra with multiplication induced naturally; in detail, if $\bar{x} \in M_i/M_{i+1}$, $\bar{y} \in M_j/M_{j+1}$, and $x \in M_i$, $y \in M_j$ have $\bar{x}$ and $\bar{y}$ as projections then $\bar{x}\cdot\bar{y} \in M_{i+j}/M_{i+j+1}$ is the projection of $x\cdot y \in M_{i+j}$. The condition $M_i \cdot M_j \in M_{i+j}$ insures that $\bar{x}\cdot\bar{y}$ is independent of choice of x and y.

<u>Filtered graded module</u>: if M is graded by $M = \sum {}^iM$ then we further require of a filtration $M = \cup M_i$ that it satisfies $M_i = \sum M_i \cap {}^jM$. It then follows that GrM is bigraded naturally.

<u>Differential filtered module</u>: a filtered A-module M with a differential d. (In many applications we have $dM_i \subset M_i$ but we do not require it here).

If M is a differential filtered module then H(M) is filtered naturally by the submodules:

$$J_i = \{x \in H(M) \mid x \text{ contains a cycle of } M_i\}.$$

Examples:

(i) A,B: differential graded modules,

$M = A \otimes B$, d the usual total differential,

$M_i = \sum_{j \geq i} {}^jA \otimes B$ (or $M_i = \sum_{j \geq i} A \otimes {}^jB$).

(ii) X : a compact complex analytic manifold, dimension n,

M : set of exterior differential forms on X,

d : exterior differentiation,

$M_i = \sum_{p \geq i} A^{p,q}$ where the $A^{p,q}$ are the modules of forms of type (p,q) defined as follows. Let $(z_1,\dots,z_n)$ be a local coordinate system on $\checkmark$; then each element of $A^{p,q}$ is expressed locally in terms of p of the dz_i and q of the $d\bar{z}_i$.

By de Rham's theorem H(M) is the cohomology ring of X .

(iii) Let L be a finite dimensional Lie algebra over K. Its universal enveloping algebra M is the quotient of its tensor algebra by the ideal generated by the elements $xy-yx-[x,y]$. An increasing filtration is obtained by taking as M^i the projection in M of $\sum_{j=0}^{i} \otimes^j L$. The Birkhoff-Witt theorem may be expressed by saying that GrM is isomorphic to the symmetric algebra over L.

9. Definition of a spectral sequence.

A spectral sequence consists of a sequence of differential modules(E_r,d_r), $r \geq r_o$,(r_o may be $-\infty$),such that $E_{r+1} = H(E_r)$ with respect to d_r.

Let k^r_{r+1} be the projection onto E_{r+1} of the cycles of E_r. Let k^r_s, $(r < s)$, be the projection onto E_s of the elements in E_r which are cycles for d_k, $r \leq k < s$, (i.e., x is in the domain of k^r_s if $d_r x = 0$, $d_{r+1}k^r_{r+1}x = 0,\dots,$ $d_{s-1}k^{s-1}_s k^{s-2}_{s-1}\cdots k^r_{r+1}x = 0$). Thus we may write $k^r_s = k^{s-1}_s k^{s-2}_{s-1}\cdots k^r_{r+1}$ for $r<s$. We put k^r_r = identity.

A cycle $x \in E_r$ is a permanent cycle if $d_i k^r_i x = 0$ for all $i \geq r$.

The limit E_∞ of a spectral sequence is defined as follows. An element of E_∞ is a sequence (x_i) where x_i is a permanent cycle of E_i and $x_{i+1} = k^i_{i+1}x_i$. Two elements (x_i) and (y_i) are identified if $x_i = y_i$ for i big enough. We make E_∞ into an A-module in the obvious way by defining

$$(x_i) + (y_i) = (x_i + y_i) \; ; \quad \alpha(x_i) = (\alpha x_i), \; a \in A.$$

The same definitions apply to spectral sequences of graded (bigraded) differential modules and differential algebras. E_∞ is then naturally graded (bigraded) and in the case of algebras E_∞ becomes an A-algebra under the definition $(x_i) \cdot (y_i) = (x_i \cdot y_i)$.

In the standard cases E_r is graded and d_r is homogeneous of degree $\pm r$ (i.e., $d_r E_r^p \subset E_r^{p\pm r}$).

A <u>canonical spectral sequence for homology</u> (E^r, d_r), $(r \geq r_o)$, has the properties:

(a) E^r is a differential module bigraded by $E^r_{p,q}$, and $E^r_{p,q} = 0$ for $p < 0$ or $q < 0$;

(b) d_r has bidegree $(-r, r-1)$.

A <u>canonical spectral sequence for cohomology</u> (E_r, d_r), $(r \geq r_o)$, has the properties

(a) E_r is a differential algebra bigraded by $E_r^{p,q}$, and $E_r^{p,q} = 0$ for $p < 0$ or $q < 0$;

(b) d_r has bidegree $(r, 1-r)$;

(c) the automorphism w_r of E_r is given by

$$w_r(x) = (-1)^{\text{total } d^o x} \cdot x$$

where x is a homogeneous element with respect to the total degree.

10. <u>Spectral sequence of a differential filtered module</u>.

We shall not make use of the material in this section in the sequel. Our objective is only to illustrate to some extent the origin and purpose of a spectral sequence.

Let M be a differential filtered module (decreasing filtration) with differential d. Then the associated graded module is $GrM = \sum M_i/M_{i+1}$. The derived module H(M) is filtered by

$$J_i = \{ x \in H(M) \mid x \text{ contains a cycle in } M_i \},$$

and its associated graded module is $GrH(M) = \sum J_i/J_{i+1}$. If we set

$$C^p = \{ x \mid x \text{ is a cycle in } M_i \},$$

$$D^p = dM \cap M_p,$$

then

$$J_i/J_{i+1} = C^p/C^{p+1} + D^p.$$

The spectral sequence connects GrM and GrH(M) by "successive approximations" as we shall now describe. Let

$$C_r^p = \{ x \mid x \in M_p,\ dx \in M_{p+r} \},$$

$$D_r^p = dC_r^{p-r} ;$$

then we have inclusions $C_r^p \supset C_{r-1}^{p+1}$ and $D_r^p \subset C_r^p$ (as well as others). We define

$$E_r^p = C_r^p \;/\; C_{r-1}^{p+1} + D_{r-1}^p,$$

$$E_r = \sum E_r^p \;;$$

then E_r is a graded module. We define a differential d_r as follows. We have

$$(C_r^p,\; C_{r-1}^{p+1} + D_{r-1}^p) \overset{d}{\to} (C_r^{p+r}, dC_{r-1}^{p+1}) \subset (C_r^{p+r}, C_{r-1}^{p+r+1} + D_{r-1}^{p+r});$$

hence we have an induced map

$$E_r^p = C_r^p \;/\; C_{r-1}^{p+1} + D_{r-1}^p \overset{d_r}{\to} C_r^{p+r} \;/\; C_{r-1}^{p+r+1} + D_{r-1}^{p+r} = E_r^{p+r}.$$

We define $d_r = \sum d_r^p$. Clearly $d_r d_r = 0$. As an exercise the reader may show that $E_{r+1} = H(E_r)$ with respect to d_r. Thus we have a spectral sequence (E_r, d_r) associated with M.

As $r \to +\infty$ one sees that $E_r^p = C_r^p \;/\; C_{r-1}^{p+1} + D_{r-1}^p$ tends to $C^p \;/\; C^{p+1} + D^p = J_i/J_{i+1}$. Thus $GrH(M) \subset E_\infty$; with suitable assumptions we obtain $GrH(M) = E_\infty$.

Similarly $E_r^p = C_r^p \;/\; C_{r-1}^{p+1} + D_{r-1}^p$ tends to M_p/M_{p+1} as $r \to -\infty$, and under suitable assumptions we get $E_{-\infty} = GrM$. Usually one assumes that $dM_i \subset M_i$, in which case we have $C_r^p = M_p$ for $r \geq 0$; then

$$C_o^p \;/\; C_o^{p+1} + D_o^p = M_p \;/\; M_{p+1},\; E_o = GrM.$$

Therefore (under suitable assumptions) the spectral sequence of M may be viewed as a sequence

$$GrM = E_o, E_1, E_2, \ldots, E_\infty = GrH(M).$$

If M is graded by submodules ${}^iM$ we define

$$C_r^{p,q} = C_r^p \cap {}^{p+q}M,\quad D_r^{p,q} = D_r^p \cap {}^{p+q}M;$$

then E_r is bigraded by submodules

$$E_r^{p,q} = C_r^{p,q} \;/\; C_{r-1}^{p+1,q-1} + D_{r-1}^{p,q}.$$

Now assume

$$(\alpha)\quad M_i = M \quad \text{for } i \leq 0,$$

$$(\beta)\quad M_i \cap {}^jM = 0 \quad \text{if } i > j.$$

(These imply that ${}^jM = 0$ for $j < 0$.) We assert that $E_r^{p,q} = 0$ if $p < 0$ or $q < 0$. Suppose first that $p < 0$, then by (α), $M_p = M = M_{p+1}$. Therefore $C_r^{p,q} = C_{r-1}^{p+1,q-1}$ and hence $E_r^{p,q} = 0$. Suppose $q < 0$, then by (β) we have $M_p \cap {}^{p+q}M = 0$. Therefore $C_r^{p,q} = 0$ and hence also $E_r^{p,q} = 0$.

Assume further that $d({}^jM) \subset {}^{j+\epsilon}M$. Then it follows easily that d_r has bidegree $(r, \epsilon - r)$. If M is also an algebra with automorphism w such that $w(M_i) \subset M_i$ it follows easily that w induces an automorphism w_r on E_r satisfying

$$d_r w_r + w_r d_r = 0,$$
$$d_r(x\cdot y) = (d_r x)\cdot y + w_r(x) d_r(y).$$

Thus E_r is a bigraded differential algebra. If we assume (α,β) as well as $\varepsilon = +1$ and $w(x) = (-1)^{d^o x}x$ it follows that (E_r, d_r) is a canonical spectral sequence for cohomology.

For $r > p$ or $r > q+1$ one sees readily that $E_r^{p,q} = E_{r+1}^{p,q} = \cdots = E_\infty^{p,q}$, and it follows that $E_\infty = GrH(M)$.

$H(M)$ is bigraded by $J^{p,q} = J_p \cap H^{p+q}(M)$, and for fixed k we have the filtration

$$H^k(M) = J^{o,k} \supset J^{1,k-1} \supset \cdots \supset J^{k,o} \supset 0$$

with the successive quotients $E_\infty^{i,k-i} = J^{i,k-i} / J^{i+1,k-i-1}$.

An alternative procedure for defining the spectral sequence of a differential filtered module is described in [11].

Let M, M' be differential filtered modules and $f : M \to M'$ a homomorphism. Then f induces homomorphisms $f_r : E_r \to E_r'$ which commute with the k_{r+1}^r. It follows that f induces a homomorphism $f_\infty : E_\infty \to E_\infty$. Note that if we regard $E_\infty \supset GrH(M)$ then f_W induced by f is the restriction of f_∞. We have the following fundamental theorem.

<u>Theorem 10.1</u>. If $f : M \to M'$ and the filtrations of M and M' are limited above (or if M and M' are graded in such a way that they determine canonical spectral sequences) and if f_{r_o} is an isomorphism for some r_o then $f^* : H(M) \to H(M')$ is an isomorphism.

The proof is easy. Clearly f_r is an isomorphism for all $r \geq r_o$. Hence f_∞ and f_W are isomorphisms and the theorem follows by lemma 8.1.

11. <u>Systems of local coefficients</u>.

Let X be a topological space and let G be some algebraic structure (e.g. a group, ring, module, or algebra). A <u>system of local coefficients</u> (G,Φ) in X consists of the following:

(a) To each $P \in X$ corresponds an isomorphism $\Psi_P : G_P \to G$,

(b) To each path homotopy class α_{PQ} joining P to Q corresponds an isomorphism $\Phi(\alpha_{PQ}) : G_P \to G_Q$ such that if $\alpha_{PR} = \alpha_{QR}\alpha_{PQ}$ then $\Phi(\alpha_{PR}) = \Phi(\alpha_{QR})\Phi(\alpha_{PQ})$.

This implies that if α_{PP} is the class of the trivial path at P then $\Phi(\alpha_{PP})$ is the identity map and also $\Phi(\alpha_{PQ}^{-1}) = \Phi(\alpha_{PQ})^{-1}$.

A <u>homomorphism</u> $f : (G,\Phi) \to (G',\Phi')$ of systems of local coefficients in X consists of a family of homomorphisms $f_P : G_P \to G_P'$ which commute with the $\Phi(\alpha_{PQ})$ and $\Phi'(\alpha_{PQ})$. If each f_P is an isomorphism then we say f is an <u>isomorphism</u>. If each f_P is a monomorphism then (G,Φ) may be identified with a sub-system of (G',Φ') the latter notion being defined in the obvious way.

Each system (G,Φ), in X clearly induces a homomorphism $\Phi_P : \pi_1(X,P) \to \mathrm{Aut}(G_P)$. If X is arcwise connected then Φ_P completely determines (G,Φ) as follows. Consider the set of all pairs (g,α_{PQ}) where $g \in G_P$ and α_{PQ} is a homotopy class of paths from P to Q. Identify $(g,\alpha_{PQ}) = (h,\beta_{PQ})$ if $h = \Phi_P(\beta_{PQ}^{-1}\alpha_{PQ})\cdot g$, and let G_Q be the set so obtained. We define $\Phi(\alpha_{PQ})$ in the obvious way, namely, $\Phi(\alpha_{PQ})\cdot g = (g,\alpha_{PQ})$.

The system (G,Φ) is said to be simple if for all $P,Q \in X$ the map $\Psi_Q\Phi(\alpha_{PQ})\Psi_P^{-1}$ is the identity map of G. When this is the case the isomorphisms Ψ_P provide a canonical identification of the G_P with G. If X is arcwise connected then (G,Φ) is simple if and only if the induced homomorphism $\Phi_P : \pi_1(X,P) \to \mathrm{Aut}(G_P)$ is trivial. In particular (G,Φ) is simple if X is simply connected. In general the set $G^{\pi_1(X)}$ of elements on which $\pi_1(X)$ acts trivially determines a subsystem $(G,\Phi)^{\pi_1(X)}$ of (G,Φ) which is actually the maximal simple subsystem. We shall usually write G^f for $G^{\pi_1(X)}$, $\tilde{G}$ for (G,Φ), and $\tilde{G}^f$ or $(G,\Phi)^f$ for $(G,\Phi)^{\pi_1(X)}$.

If X is a locally finite polyhedron then we can give a simpler definition for (G,Φ):

(a) To each vertex P corresponds an isomorphism $\Psi_P : G_P \to G$,

(b) If P and Q are vertices of a simplex then there is an isomorphism $\Phi_{PQ} : G_P \to G_Q$ such that if P,Q,R are in a simplex then $\Phi_{PR} = \Phi_{QR}\cdot\Phi_{PQ}$.

This is essentially equivalent to the original definition. For if P and Q are vertices of X then to each equivalence class of routes α_{PQ} joining P to Q we define $\Phi(\alpha_{PQ})$ as the composition of the isomorphisms corresponding to adjacent vertices of a representative route in α_{PQ}. If P and Q are not vertices and α_{PQ} is a path homotopy class joining them then we can define $\Phi(\alpha_{PQ})$ by making use of a simplicial approximation.

We shall assume hereafter that G is abelian.

Cohomology with local coefficients:

$C^p(X,\tilde{G})$ consists of functions f defined on (p+1)-tuples $(P_0,\ldots,P_p)$ of vertices in X with values $f(P_0,\ldots,P_p) \in G_{P_0}$. We define the coboundary δf by the formula

$$\delta f(P_0,\ldots,P_{p+1}) = \Phi_{P_1P_0} f(P_1,\ldots,P_{p+1}) + \sum (-1)^i f(P_0,\ldots,\hat{P}_i,\ldots,P_{p+1}).$$

In the usual way one shows $\delta\delta = 0$, and we define

$$H^p(X,\tilde{G}) = \delta^{-1}(0) \;/\; \delta C^{p-1}(X,\tilde{G}).$$

If X is connected then $H^o(X,\tilde{G}) = G^f$. For if f is a o-cocycle then

$$\delta f(P_oP_1) = \Phi_{P_1P_o} f(P_1)-f(P_o) = 0.$$

Thus $\pi_1(x)$ acts trivally on $f(P)$, and conversely.

By a pairing μ: $\tilde{G}_1,\tilde{G}_2 \to \tilde{G}_3$ we mean a family of pairings $\mu_P : G_{1P},G_{2P} \to G_{3P}$ which commute with the maps $\Phi_1(\alpha_{PQ}) \times \Phi_2(\alpha_{PQ})$ and $\Phi_3(\alpha_{PQ})$. Given such a pairing we define a pairing

$$\mu' : C^P(X,\tilde{G}_1),C^q(X,\tilde{G}_2)\to C^{p+q}(X,\tilde{G}_3)$$

by

$$\mu'(fg)\ (P_o,\ldots,P_{p+q}) = \mu'_{P_o}\ (f(P_o,\ldots,P_p),\ \ \Phi_{P_pP_o}\ g(P_p,\ldots,P_{p+q})).$$

As usual the coboundary formula $\delta fg = \delta f\cdot g + (-1)^p f\cdot\delta f$ is satisfied. Thus we can define a pairing

$$\mu^* : H^p(X,\tilde{G}_1),\ H^q(X,\tilde{G}_2) \to H^{p+q}(X,\tilde{G}_3)$$

in the usual way. If $\tilde{G}_1 \approx \tilde{G}_2 \approx \tilde{G}_3$ are simple μ^* reduces to the ordinary cup product.

Homology with local coefficients:

$C_p(X,\tilde{G})$ consists of finite linear combinations of p+1-tuples of vertices of X such that $(P_o,\ldots,P_p)$ has a coefficient in G_{P_o}. The boundary is defined by the formula

$$\partial g\ (P_o,\ldots,P_p) = \Phi_{P_1P_o} g\cdot(P_1,\ldots,P_p) + \sum (-1)^i g\cdot(P_o,\ldots,\hat{P}_i,\ldots,P_p).$$

As usual we have $\partial\partial = 0$ so that we can define

$$H_p(X,\tilde{G}) = \partial^{-1}(0)\ /\ \partial C_{p+1}(X,\tilde{G}).$$

If X is connected one verifies readily that

$$H_o(X,\tilde{G}) = G\ /\ [g-\gamma g],$$

where γ is an automorphism of G defined by $\pi_1(X)$ and $[g-\gamma g]$ is generated by elements of the form $g-\gamma g$.

12. Fibre bundles.

A (locally trivial) fibre bundle consists of three topological spaces E,B,F and a continuous surjective map $\pi : E \to B$ subject to the following condition: for each $b \in B$ there exists a neighborhood V_b of b and a homeomorphism Ψ_b: $\pi^{-1}(V_b) \to V_b \times F$ such that $\Psi_b(\pi^{-1}(b)) = b \times F$. We shall denote the fibre bundle by (E,B,F,π) or (E,B,F): the spaces E,B,F are called respectively the total space, base space, and standard fibre and π is called the projection map. Sometimes we shall refer to E as the fibre bundle. For $b \in B$, $\pi^{-1}(b)$ is called the fibre over b : it is evidently homeomorphic to the standard fibre F.

A representation (also homomorphism) of (E,B,F,π) in (E',B',F',π') is a map

$\Phi : E \to E'$ which maps the fibres over the points of B into fibres over points of B'. In other words Φ induces a map $\bar{\Phi} : B \to B'$ such that $\pi'\Phi = \bar{\Phi}\pi$. If in addition Φ is a homeomorphism then we say (E,B,F,π) is <u>isomorphic</u> to (E',B',F',π'). A fibre bundle (E,B,F,π) is <u>equivalent</u> to (E',B',F,π') (same base and fibre) if there is a representation of (E,B,F,π) in (E',B,F,π') such that the induced map $\bar{\Phi} : B \to B$ is the identity.

Let G be a topological group operating on F (with the usual continuity conditions but not necessarily effectively). We say that the bundle has <u>structural group</u> G if the following holds: there is a covering of B by open sets U_i and a corresponding family of continuous maps $g_{ij} : U_i \cap U_j \to G$ with the properties:

(a) $g_{ii} = \mathrm{id}$, $g_{ij}g_{ji} = \mathrm{id}$,

(b) $g_{ij}g_{jk}g_{ki} = \mathrm{id}$ in $U_i \cap U_j \cap U_k$,

(c) The homeomorphism $\psi_i \psi_j^{-1}$ of $\psi_j(\pi^{-1}(U_i \cap U_j))$ (in the above notations) is given by $(b,f) \to (b,g_{ij}(b)f)$.

It is of course always true that in the case of a local trivial fibre bundle we can write $\psi_i\psi_j^{-1}$ in the form $(b,f) \to (b,g_{ij}(b)f)$ with $g_{ij}(b) \in \mathrm{Aut}\ F$ and satisfying also conditions (a) and (b). It can be shown further that the required continuity conditions hold if F is locally compact and Aut F is given the compact-open topology with respect to g and g^{-1}, i.e., the topology defined by the sets

$$(C,V) = \{g \in \mathrm{Aut}\ F \mid g(C) \subset V,\ g^{-1}(C) \subset V\}$$

where C is a compact set and V is an open set of F.

A <u>cross-section</u> of (E,B,F) is a map $s : B \to E$ such that $\pi \cdot s$ = identity. We shall also speak of the image of s as a cross-section. By a <u>local cross</u>-section we mean a map $s : B \to E$ which is a cross-section in a neighborhood of every point. Clearly local cross-sections exist.

A fibre bundle which is equivalent to $B \times F$ is called a <u>trivial bundle</u>.

Let (E,B,F,π) be a fibre bundle and let $\Phi : B' \to B$. An <u>induced bundle</u> for Φ is a fibre bundle (E',B',F,π') for which there is a representation in (E,B,F,π) with Φ as the induced map on B'. Define $E' \subset B' \times E$ as the subset of pairs (b',e) such that $\Phi(b') = \pi(e)$. Let π' and $\tilde{\Phi}$ be defined by $(b',e) \to b'$ and $(b',e) \to e$ respectively. Then (E',B',F,π') is an induced bundle for Φ. For (locally trivial) fibre bundles every induced bundle for Φ is equivalent to (E',B',F,π').

E is a <u>principal bundle</u> with structural group a topological group G if G operates on the right on E as follows:

(a) $(e \cdot g) \cdot g' = e \cdot gg'$, $e \cdot 1 = e$,

(b) $e \cdot g = e$ for some e implies that $g = 1$,

(c) On the subspace of pairs (e,e') for which there is a g such that $e \cdot g = e'$ this unique g is a continuous function of e and e'.

For fixed e its <u>orbit</u> is the set of all $e \cdot g$, $g \in G$. Conditions (a,b,c) imply that the orbits are closed subspaces of E which are homeomorphic to G. If G is compact then (c) can be readily shown to be a consequence of (a,b), but in general this is not true as the following example shows:

Let E be the 2-dimensional torus and regard it as the quotient space of the Euclidean plane modulo the lattice of points with integer coefficients. Let L be a line through the origin with irrational slope. Then L determines a 1-parameter group G which operates on E by right translations. Note that G is not compact. The orbit through e is evidently dense in E and hence is not closed. Since (a,b) clearly hold, (c) must fail.

Let $B = E/G$ denote the space of orbits with the natural topology and let $\pi : E \to B$ be the natural projection. The fibres F_b are precisely the orbits of E. By a theorem of A.M. Gleason, if E is regular and G is a Lie group then (E,B,π) is a (locally trivial) fibre bundle with G as structural group. Moreover, in the case where G is locally compact and $B = E/G$ is locally contractible it can be shown that (E,B,π) is a (locally trivial) fibre bundle with G as structural group. The following example shows however that in general (E,B,π) is not locally trivial:

Let E be the product of infinitely many copies of SO(3). Let $H \subset SO(3)$ be a 1-parameter subgroup, and let G be the corresponding infinite product of H. Then G is a closed subgroup of E. Suppose there existed a local cross-section for (E,B,π). Then from the definition of the product topology it is clear that this would imply that $SO(3) \to SO(3)/H$ has a cross-section. But it is known that no such cross-sections exist, and hence (E,B,π) is not locally trivial.

The following are examples of principal bundles:

(i) E a topological group, G a closed subgroup operating on E by right translations. B is then the space of left cosets.

(ii) E the space of frames on a differentiable manifold M of dimension n, G the general linear group of order n. The base space B is then M itself.

(ii)' E the space of orthonormal frames on a Riemannian manifold of dimension n, G the orthogonal group O(n).

Let (E,B,G,π), (E',B',G',π') be two principal bundles. We say that a map $\Phi : E \to E'$ is a <u>representation</u> (or a homomorphism) of (E,B,G) into (E',B',G') if there is a homomorphism $\Psi : G \to G'$ such that $\Phi\,(e\cdot g) = \Phi(e)\cdot\Psi(g)$. Then Φ maps fibers into fibers and induces a map $\bar{\Phi} : B \to B'$ such that $\bar{\Phi}\cdot\pi = \pi'\cdot\Phi$. The representation is an <u>isomorphism</u> if Φ is a homeomorphism. We say (E,B,G,π) is trivial if it is isomorphic to $B \times G$, ($\bar{\Phi}$ being the identity), the latter being acted upon by G in the natural way. By a well known theorem, a principal bundle is trivial if and only if it has a cross-section.

Let E be a principal bundle with structural group G. Let F be a topological space on which G operates on the right. We consider $E \times F$ and define (e,f) and (e',f') equivalent if there is a $g \in G$ such that $e' = e\cdot g$ and $f' = f\cdot g$. Denote the quotient space by $(E,F)_G$. We define $\bar{\pi} : (E,F)_G \to B$ by the diagram

$$\begin{array}{ccc} E \times F & \longrightarrow & E \\ \downarrow & & \downarrow \pi \\ (E,F)_G & \xrightarrow{\ \bar{\pi}\ } & B \end{array}$$

where the unlabelled maps are the natural projections. It is easy to verify that $((E,F)_G,B,F,\bar{\pi})$ is a fibre bundle. For example to show that the standard fibre is F choose a fixed point e_o in $\pi^{-1}(b)$. Then any pair (e,f) is equivalent to a pair (e_o,fg^{-1}) for some g; hence each element of $(E,F)_G$ contains a pair of the form (e_o,f), and moreover, it is easy to see that it can contain no other such pair. This shows that $\bar{\pi}^{-1}(b)$ is in 1-1 correspondence with F. We leave the remaining details to the reader.

The bundle $((E,F)_G,B,F,\bar{\pi})$ is called an <u>associated bundle</u> of E. If the principal bundle E is locally trivial so are its associated bundles. It can be shown that the associated bundles are fibre bundles with structural groups. Moreover, every (locally trivial) fibre bundle with a structural group is obtainable from a principal bundle by the above process. For the details we refer the reader to [13] (section 8) or [4].

Let E_1, E_2 be principal bundles with the same group G. By means of the diagram

$$\begin{array}{ccccc} E_1 & \longleftarrow & E_1 \times E_2 & \longrightarrow & E_2 \\ \pi_1 \downarrow & & \downarrow & & \downarrow \pi_2 \\ B_1 & \xleftarrow{\ \bar{\pi}_1\ } & (E_1,E_2)_G & \xrightarrow{\ \bar{\pi}_2\ } & B_2 \end{array}$$

in which the unlabelled maps are natural projections, we obtain by the same process as above a pair of associated bundles,

$$((E_1,E_2)_G, B_2, E_1, \bar{\pi}_1), \quad ((E_1E_2)_G, B_2E_1, \pi_2).$$

(The above constructions are due to C. Ehresmann.)

Let (E,B,F,π) be a fibre bundle with B a locally finite polyhedron. Assuming a sufficiently fine simplicial subdivision we construct a system of local coefficients $\tilde{H}^*(F,A) = (H^*(F,A),\phi^*)$ as follows. If P is a vertex of B then on StP, (star of P), the bundle is trivial. Thus we have a homeomorphism

$$\Psi_P : \pi^{-1}(StP) \to StP \times F,$$

and hence a homeomorphism

$$\psi_P : F_P \to P \times F$$

where F_P is the fibre over P. This defines an isomorphism

$$\psi_P^* : H^*(F,A) \to H^*(F_P,A).$$

If P,Q are vertices of a simplex then $U_{PQ} = StP \cap StQ$ is not empty. On $U_{PQ} \times F$ we define $f_{PQ} = \Psi_P \cdot \Psi_Q^{-1}$; this determines an isomorphism

$$f_{PQ}^* : H^*(F,A) \to H^*(F,A).$$

We define

$$\phi_{PQ}^* : H^*(F_P,A) \to H^*(F_Q,A)$$

by $\phi_{PQ}^* = \psi_{PQ}^* f_{PQ}^* (\psi_{PQ}^*)^{-1}$. It is easily seen that we have a system of local coefficients. Similarly we can define a system $\tilde{H}_*(F,A)$, and if F is simple, systems consisting of homotopy groups.

13. Spectral sequences for fibre bundles

Theorem 13.1. Let (E,B,F,π) be a fibre bundle with E,B, and F connected locally finite polyhedra. Let L be a principal ideal ring and A an L-module. Then there is a filtration of $H_*(E,A)$ and a canonical spectral sequence for homology, $(E^r,dr), r \geq 2$, such that

$$E^\infty \approx GrH_*(E,A),$$
$$E^2_{p,q} \approx H_p\ (B,\tilde{H}_q(F,A)),$$

where $\tilde{H}_q(F,A)$ denotes the system $(H_*(F_b,A),\phi_*)$.

Note that E^r is an L-module.

Theorem 13.2. Let (E,B,F,π) be a fibre bundle with E,B, and F connected locally finite polyhedra. Let L be a principal ideal ring and A an L-algebra. Then there is a filtration of $H^*(E,A)$ and a canonical spectral sequence for cohomology, (E_r,d_r), $r \geq 2$, such that

$$E_\infty \approx \mathrm{Gr}H^*(E,A),$$
$$E_2^{p,q} \approx H^p(B,\tilde{H}^q(F,A)),$$

where $\tilde{H}^q(F,A)$ denotes the system $(H^*(F_b,A),\phi^*)$. Moreover, if A is commutative then E_r is an L-algebra which is anticommutative with respect to the total degree, and the isomorphism

$$E_2 \approx H^*(B,\tilde{H}^*(F,A))$$

is multiplicative.

We remark that the theorems have been proved under the following other assumptions:

(i) E,B,F locally compact, Čech cohomology with compact carriers ([1],[9]).

(ii) E,B,F locally compact, E a principal bundle with connected compact structural group, Čech cohomology with compact carriers [9].

(iii) E,B,F arbitrary, (E,B,F) a "fibre space" in the sense of Serre, singular homology (and cohomology) [12].

(iv) B paracompact, F compact, Čech cohomology with arbitrary closed carriers [5].

(v) E,B,F locally finite polyhedra ([6],[7],[10]).

The last is a special case of (iii). We note further that for real cohomology of differentiable bundles the spectral sequence may be constructed using exterior differential forms (provided F is compact).

We list the main properties of the spectral sequence for cohomology. In the applications it is these rather than the actual definitions which are most useful.

(13.3) If $\tilde{H}^*(F,A)$ is a simple system, as is the case if B is simply connected or the structural group is connected, then we may canonically identify $H^*(F_b,A)$ with $H^*(F,A)$ by the isomorphism ψ_b^*. Then since the cohomology groups are finitely generated we have, by the Künneth rule,

$$0 \to H^p(B,A) \otimes H^q(F,A) \to E_2^{p,q} \to \mathrm{Tor}\,(H^{p+1}(B,A),H^q(F,A)) \to 0$$

where the tensor and torsion products are taken over L. Thus in particular if A is a field K we have

$$E_2^{p,q} \approx H^p(B,K) \otimes H^q(F,K).$$

Similarly, if A = Z and if B or F has no torsion then

$$E_2^{p,q} \approx H^p(B,Z) \otimes H^q(F,Z).$$

(13.4) Clearly $d_r E_r^{p,o} = 0$; hence

$$k_{r+1}^r : E_r^{p,o} \to E_{r+1}^{p,o},\ r \geq 2$$

is surjective so that

$$E_2^{p,o} \xrightarrow{k_3^2} E_3^{p,o} \xrightarrow{k_4^3} \cdots \xrightarrow{k_{p+1}^p} E_{p+1}^{p,o} = E_\infty^{p,o}$$

is also surjective. Recall that

$$E_\infty^{p,o} = J^{p,o} \subset H^p(E,A),$$

$$E_2^{p,o} \approx H^p(B,\tilde{H}^o(F,A)) = H^p(B,A).$$

Then (13.4) asserts that π^* is identical with the composed map

$$H^p(B,A) \approx E_2^{p,o} \xrightarrow{k_{p+1}^2} E_\infty^{p,o} \approx J^{p,o} \subset H^p(E,A).$$

(13.5) From $E_r^{p,q} = 0$ for $p < 0$ it follows that $E_{r+1}^{p,q}$ is the submodule of d_2-cocycles of $E_r^{p,q}$. We therefore have

$$E_\infty^{o,q} = E_{q+2}^{o,q} \subset E_{q+1}^{o,q} \subset \cdots \subset E_2^{o,q}.$$

Also

$$E_2^{o,q} \approx H^o(B,\tilde{H}^q(F,A)) = H^q(F_b,A)^f \subset H^q(F_b,A),$$

(recall $H^q(F_b,A)^f$ consists of the elements fixed under the action of $\pi_1(B)$.) We have $E_\infty^{o,q} \approx H^q(E,A)/J^{1,q-1}$. Then (13.5) asserts that the map i_b^* given by the inclusion $F_b \subset E$ is identical with the composed map

$$H^q(E,A) \to E_\infty^{o,q} \subset E_2^{o,q} \approx H^q(F_b,A)^f \subset H^q(F_b,A).$$

The kernel of i_b^* is J_1, the image of i_b^* is the submodule of permanent cocycles which are fixed under the action of $\pi_1(B)$. The isomorphisms $\psi_b^* : H^*(F,A) \to H^*(F_b,A)$ give a canonical identification of $H^q(F_b,A)^f$ with the "fixed" submodule $H^q(F,A)^f \subset H^q(F,A)$. Thus we may identify all i_b^* with

$$i^*: H^q(E,A) \to H^q(F,A)^f \subset H^q(F,A),\ i^* = (\psi_b^*)^{-1} i_b^*.$$

(13.6) The map $d_{q+1} : E_{q+1}^{o,q} \to E_{q+1}^{q+1,0}$ is called the <u>transgression</u>. Let $T^q(F,A)$ be the submodule of $H^q(F,A)$ which corresponds under the isomorphism $E_2^{o,q} \approx H^q(F,A)^f$ to $E_{q+1}^{o,q}$. We saw above that $E_{q+1}^{q+1,0}$ is isomorphic to a factor module of $H^{q+1}(B,A)$ by some submodule which we shall denote by M^{q+1}. The diagram

$$\begin{array}{ccc} E_{q+1}^{o,q} & \xrightarrow{d_{q+1}} & E_{q+1}^{q+1,0} \\ \approx \downarrow & & \downarrow \approx \\ T^q(F,A) & \xrightarrow{\tau} & H^{q+1}(B,A)/M^{q+1} \end{array}$$

defines the map τ which we also call the <u>transgression</u>. The elements of $T^q(F,A)$ are said to be <u>transgressive</u>.

There are various other definitions of the transgression, see [2]; the following one is very useful:

The maps $i_b : F_b \subset E$ and $\pi : E \to B$ induce cochain maps

$$i' : C^q(E,A) \to C^q(F,A), \quad \pi' : C^q(B,A) \to C^q(E,A).$$

An element $x \in H^q(F,A)$ is said to be <u>transgressive</u> if there exists a cochain $e \in C^q(E,A)$ such that

(a) $i'e$ is a cocycle in x,

(b) $\delta e = \pi'(b)$, where $b \in C^{q+1}(B,A)$.

Since π' is injective this implies that b is a cocycle. If y is the cohomology class of b then we write $\tau(x) = y$. This defines the <u>transgression</u> τ. This definition can be proved to be equivalent to the preceding one.

(13.7) Let (E,B,F,π) and (E', B',F',π') be fibre bundles with simple systems $\tilde{H}^*(F,A)$ and $\tilde{H}^*(F',A)$ respectively. Let

$$\begin{array}{ccc} E & \xrightarrow{\lambda} & E' \\ \pi \downarrow & & \downarrow \pi' \\ B & \xrightarrow{\bar{\lambda}} & B' \end{array}$$

be a representation; then there is induced a homomorphism of spectral sequences, $\lambda^*: (E'_r) \to (E_r)$. The induced map $\lambda^* : H^*(E',A) \to H^*(E,A)$ satisfies $\lambda^* (J'^{p,q}) \subset J^{p,q}$, and hence defines a map

$$\lambda^*_W : GrH^*(E',A) \to GrH^*(E,A).$$

Let π_b be the restriction of π to the fibre F_b; then π_b together with $\bar{\lambda}$ induces a map

$$H^p(B',H^q(F',A)) \to H^p(B,H^q(F,A)).$$

Applying the isomorphism $E_2^{p,q} \approx H^p(B,H^q(F,A))$ we obtain maps $\lambda_2^* : E'^{p,q}_2 \to E_2^{p,q}$, $\lambda_r^* : E'^{p,q}_r \to E_r^{p,q}$ $(r \geq 2)$, and eventually a map $\lambda^*_\infty: E'^{p,q}_\infty \to E^{p,q}_\infty$. The maps λ^*_∞ and the above map λ^*_W are identical.

(13.8) Let F be a finite polyhedron of dimension n and let (E,B,F,π) and (E',B',F,π') be fibre bundles over F with simple local systems $\tilde{H}^*(F,A)$. If λ is a representation

$$\begin{array}{ccc} E & \xrightarrow{\lambda} & E' \\ \pi \downarrow & & \downarrow \pi' \\ B & \xrightarrow{\bar{\lambda}} & B' \end{array}$$

such that λ is a homeomorphism on each fibre and $\bar{\lambda}^*$ is an isomorphism up to dimension k then the induced homomorphism $\lambda^*: (E'_r) \to (E_r)$ of the spectral sequences and the induced map $\lambda^* : H^*(E',A) \to H^*(E,A)$ are isomorphisms up to total degree k-n-1.

Proof. $\bar{\lambda}^*$ induces an isomorphism

$$\lambda^* : H^*(B',H^*(F,A)) \to H^*(B,H^*(F,A))$$

up to dimension k. In view of the isomorphisms

$$E'_2 \approx H^*(B',H^*(F,A)), \quad E_2 \approx H^*(B,H^*(F,A))$$

we see that $d_2 : E'_2 \to E_2$ is an isomorphism up to total degree k. Similarly $d_3 : E'_3 \to E_3$ is an isomorphism up to total degree k-1, and so on. However since F has dimension n it follows $d_r \equiv 0$ for $r \geq 2$ so that $E'_{n+2} = E'_\infty$ and $E_{n+2} = E_\infty$. Then $\lambda^* : E' \to E$ is an isomorphism up to total degree k-n-1, and (13.8) follows readily.

14. Some simple applications.

We call the reader's attention particularly to [9], [12], Exposé IX in [1], and Exposé X in [5].

The Betti numbers of the spectral sequence are $p_s(E_r) = \dim {}^sE_r$, $r \leq \infty$. The Poincaré polynomials and Euler characteristics will be taken with respect to the total degree; explicitly,

$$P(E_r,t) = \sum p_s t^s, \quad X(E_r) = P(E_r,-1).$$

The fibre F is said to be totally non-homologous to zero (relative to A) if the induced map $i^* : H^*(E,A) \to H^*(F,A)$ is surjective.

Theorem 14.1. If A = K then the following two conditions are equivalent:

(a) F is totally non-homologous to 0,

(b) $\tilde{H}^*(F,A)$ is simple and $E_2 = E_\infty$, (explicitly, $d_r = 0$, $r \geq 2$).

Proof. First assume (b). Referring to (13.5) we see at once that $E_2 = E_\infty$ implies that the image of i^* is $H^q(F,K)^f$. Then since $\tilde{H}^*(F,K)$ is simple it follows that the image of i^* is $H^q(F,K)$. This proves (a).

Now assume (a). Again referring to (13.5) we see that

$$i^* H^q(E,K) = E_\infty^{0,q} \subset H^q(F,K)^f \subset H^q(F,K).$$

Since by assumption i^* is surjective equality holds. Therefore $\tilde{H}^*(F,K)$ is simple and, moreover, $H^q(F,K)$ consists of permanent cocycles. Therefore $d_r(E_r^{0,q}) = 0$ for $r \geq 2$. Also observe that we always have $d_r(E_r^{p,0}) = 0$. To complete (b) it remains to show that $d_r \equiv 0$ for $r \geq 2$. Since A = K is a field we have by (13.3)

$$E_2^{p,q} \approx H^p(B,K) \otimes H^q(F,K).$$

We may thus regard an element $x \in E_2^{p,q}$ as a finite linear combination of elements $b \otimes f$, $b \in H^p(B,K), f \in H^q(F,K)$. Since the above isomorphism is multiplicative we may write each

$$b \otimes f = (b \otimes 1) \cdot (1 \otimes f)$$

and regard $b \otimes 1$ in $E_2^{p,o}$ and $1 \otimes f$ in $E_2^{o,q}$. But we have observed above that $d_2(E_2^{p,o}) = 0$ and $d_2(E_2^{o,q}) = 0$; hence it follows that $d_2\,(b \otimes f) = 0$. Thus we have shown that $d_2 \equiv 0$. Similarly we can show by induction that $d_r \equiv 0$ for $r > 2$, and it follows that $E_2 = E_\infty$. This completes the proof of (b).

Theorem 14.2. If $A = K$ and F is totally non-homologous to zero then

(a) π^* is injective,

(b) $H^*(F,K)$ is isomorphic to a factor module of $H^*(E,K)$ by the ideal generated by $\pi^* H_+^*(B,K)$, where $H_+^*(B,K) = \sum_{i \geq 1} H^i(B,K)$,

(c) $P_K(E,t) = P_K(B,t) \cdot P_K(F,t)$.

Proof (a) Referring to (13.4) we see that

$$\pi^* H^p(B,K) = J^{p,o} \subset H^p(E,K).$$

But $J^{p,o} = E_\infty^{p,o}$ which by assumption is equal to $E_2^{p,o} \approx H^p(B,K)$. Thus π^* is injective.

(b) i^* is surjective and by (13.5) has kernel J_1. Consider

$$\pi^* H^p(B,K) \cdot H^*(E,K) = J^{p,o} \cdot J_o \subset J_p \subset J_1.$$

This implies that $\pi^* H_+^*(B,K) \subset J_1$ so that the ideal $(\pi^* H_+^*(B,K))$ is contained in J_1. It remains to show that $J_1 \subset (\pi^* H_+^*(B,K))$.

Consider $J^{p-1,1} \supset J^{p,o} = \pi^* H^p(B,K)$.

We have

$$J^{p-1,1}/J^{p,o} = E_\infty^{p-1,1} = E_2^{p-1,1}$$

by our assumption. Therefore

$$J^{p-1,1}/J^{p,o} \approx H^{p-1}(B,K) \otimes H^1(F,K).$$

Let $x \in J_{p-1}$; then we may write its projection $\bar{x} \in J^{p-1,1}/J^{p,o}$ as a finite linear combination $\bar{x} = \sum \bar{b}_i \otimes \bar{f}_i$, where $\bar{b}_i \in H^{p-1}(B,K)$ and $\bar{f}_i \in H^1(F,K)$. Therefore there are elements $b_i \in H^{p-1}(B,K)$, $f_i \in H^1(F,K)$ such that $x - \sum b_i \cdot f_i \in J^{p,o}$. Since the sum is clearly in the ideal $(\pi^* H_+^*(B,K))$, (as is also $J^{p,o}$), it follows that x belongs to the ideal. Thus we have shown that

$$J^{p-1,1} \subset (\pi^* H_+^*(B,K))$$

for $p-1 > 1$, (if $p = 0$ there is nothing to prove). Similarly by induction we may show that

$$J^{i,p-i} \subset (\pi^* H_+^*(B,K)),\ i \geq 1.$$

For $i = 1$ we have

$$J^{1,p-1} \subset (\pi^* H_+^*(B,K))$$

from which it follows that J_1 is contained in the ideal $(\pi^* H_+^*(B,K))$. This completes the proof of (b).

(c) We have $E_\infty = E_2$ by (14.1); then (c) follows at once from the fact that $P_K(E,t) = P_K(E_\infty,t) = P_K(E_2,t)$.

Theorem 14.3. If $\tilde{H}^*(F,K)$ is simple then

(a) the Betti numbers satisfy $p_k(E) \leq p_k(B \times F)$,

(b) the Euler characteristics satisfy $\chi(E) = \chi(B) \cdot \chi(F)$.

Proof (a) We have

$$E_2 = H^*(B,K) \otimes H^*(F,K).$$

This implies that

$$p_k(E_2) = \sum_{i+j=k} \dim E_2^{i,j} = \sum_{i+j=k} p_i(B) \cdot p_j(F) = p_k(B \times F).$$

Since E_3 is a sub-quotient of E_2 we have

$$p_k(E_3) \leq p_k(E_2) = p_k(B \times F),$$

and more generally

$$p_k(E_r) \leq p_k(E_2) = p_k(B \times F), \quad \text{for } r \geq 2.$$

Therefore $p_k(E_\infty) \leq p_k(B \times F)$. Since $p_k(E) = p_k(E_\infty)$ (a) is proved.

(b) From the above isomorphism it follows that $\chi(E_2) = \chi(B) \cdot \chi(F)$. Since d_r increases the total degree by 1 it follows from a general theorem that with respect to the total degree

$$\chi(E_2) = \chi(E_3) = \ldots = \chi(E_\infty).$$

Since $\chi(E) = \chi(E_\infty)$, (2) is proved.

Theorem 14.4. If the Betti numbers satisfy $p_k(E) \geq p_k(B \times F)$ for all k then $H^*(F,K)$ is simple, i^* is surjective, and $p_k(E) = p_k(B \times F)$.

Proof. Clearly $\tilde{H}^0(F,K) = K$ is a simple system. Now assume that $\tilde{H}^k(F,K)$ is a simple system for $i \leq k$; we shall prove that $\tilde{H}^{k+1}(F,K)$ is also a simple system. As we have observed in the proof of (14.3)

$$p_{k+1}(E) = p_{k+1}(E_\infty) \leq p_{k+1}(E_2).$$

By the inductive assumption we have

$$E_2^{i,j} = H^i(B,K) \otimes H^j(F,K), \quad j \leq k;$$

hence

$$p_{k+1}(E_2) = \sum_{i+j=k+1} \dim E_2^{i,j} = \sum_{i+j=k+1} p_i(B)p_j(F) + \dim E_2^{o,k+1}$$

$$= \sum_{i+j=k+1} p_i(B)p_j(F) - \dim H^{k+1}(F,K) + \dim E_2^{o,k+1}$$

$$= p_{k+1}(B \times F) - \dim H^{k+1}(F,K) + \dim E_2^{o,k+1}.$$

Therefore

$$p_{k+1}(E) \leq p_{k+1}(B \times F) - \dim H^{k+1}(F,K) + \dim E_2^{o,k+1},$$
$$\leq p_{k+1}(E) - \dim H^{k+1}(F,K) + \dim E_2^{o,k+1}.$$

This implies that

$$\dim H^{k+1}(F,K) \leq \dim E_2^{o,k+1} = \dim H^{k+1}(F,K)^f,$$

and hence equality must hold. But $H^{k+1}(F,K)^f \subset H^{k+1}(F,K)$; therefore $H^{k+1}(F,K)^f = H^{k+1}(F,K)$ so that $\tilde{H}^{k+1}(F,K)$ is a simple system. Thus we have proved that $\tilde{H}^*(F,K)$ is a simple system. Then

$$E_2 \approx H^*(B,K) \otimes H^*(F,K)$$

and $p_k(E_2) = p_k(B \times F)$ for all k, whence

$$p_k(E) = p_k(E_\infty) \leq p_k(E_2) = p_k(B \times F) \leq p_k(E),$$

so that equality holds throughout. This proves that $p_k(E) = p_k(B \times F)$ and $p_k(E_\infty) = p_k(E_2)$ for all k. Hence $E_\infty = E_2$. Applying theorem (14.2) it follows that i^* is surjective.

Theorem 14.5. Let $H^i(F,K) = 0$ for $i > s$ and let $\pi_1(B)$ act trivially on $H^s(F,K)$. If there exists a non-zero element $x \in H^s(F,K)$ which is in the image of i^* then π^* is injective.

Proof. Referring to (13.4) we see that π^* is injective if and only if $(d_r E_r) \cap E_r^{p,o} = 0$ for all $r \geq 2$. Suppose that π^* is not injective. Then for some $r \geq 2$ we can find an element $b \in H^*(B,K)$, $b \neq 0$, and an element $y \neq 0$, $y \in E_r$ such that $d_r y = k_r^2 b$. Assume this is the first such r; we shall produce a contradiction.

Let $x \in H^s(F,K)$ be a non-zero element in the image of i^*, and consider $y \cdot k_r^2 x$. We have

$$d_{r_o}(y \cdot k_r^2 x) = (d_r y)(k_r^2 x) + y \cdot d_r(k_r^2 x).$$

Since x is in the image of i^* it is a permanent cocycle so that the second term vanishes. Therefore

(α) $$d_r(y \cdot k_r^2 x) = (k_r^2 b)(k_r^2 b)(k_r^2 x) = k_r^2(b \cdot x).$$

Since $d_r y = k_r^2 b$ the element y has bidegree (r,r-1) and $y \cdot k_r^2 x$ is in $E_r^{p-r,r-1+s}$.

But $r-1+s>s$ for $r\geq 2$; hence $y\cdot k_r^2 x = 0$. On the other hand since $\pi_1(B)$ acts trivially on $H^s(F)$ we have

$$E_2^{p,s} \approx H^p(B,K) \otimes H^s(F,K).$$

Under this isomorphism $b\cdot x$ corresponds to $b \otimes x$ and is thus not 0. Clearly $b\cdot x$ is a permanent cocycle, and having maximal fibre degree it cannot be a coboundary. But then $k_r^2(b\cdot x) \neq 0$, which by (α) contradicts $y\cdot k_r^2 x = 0$.

The above theorems were proved under the assumption that A was a field K. If $H^*(B)$ or $H^*(F)$ has no torsion the theorems remain valid if we take $A = Z$.

If i^* is surjective then by theorem (14.2) π^* is injective. The converse however is false as is shown by the following example (due to G. Hirsch) which we shall outline briefly:

Consider the Hopf fibering $S_7 \to S_4$ with fibre S_3. We have $E_2 = H^*(S_4) \otimes H^*(S_3)$; E_2 is thus determined by generators y and x of degrees 4 and 3, respectively, with $d_4 y = x$. By a well-known result there is a map

$$f : S_2 \times S_2 \to S_4$$

of degree 1. Consider the bundle $(E, S_2 \times S_2, S_3)$ induced by f,

$$\begin{array}{ccc} E & & S_7 \\ \downarrow{\scriptstyle \pi'} & & \downarrow{\scriptstyle \pi} \\ S_2 \times S_2 & \xrightarrow{f} & S_4 \end{array}$$

In this bundle $d_4 \neq 0$ so that $E_2 \neq E_\infty$. It follows that

$$P(E,t) \neq P(S_2 \times S_2 \times S_3, t).$$

Now consider a new bundle $(E, S_2, S_2 \times S_3, \pi''\cdot\pi')$ determined by

$$\begin{array}{ccc} E & & S_7 \\ \downarrow{\scriptstyle \pi'} & & \downarrow{\scriptstyle \pi} \\ S_2 \times S_2 & \xrightarrow{f} & S_4 \\ \downarrow{\scriptstyle \pi''} & & \\ S_2 & & \end{array}$$

where π'' is the projection on one of the factors. The fiber is a principal bundle which has a cross-section and is therefore homeomorphic to $S_2 \times S_3$. On the other hand $(\pi''\cdot\pi')^*$ is clearly injective (and in fact there is a cross section), but the injection map i^* cannot be surjective in this bundle since

$$P(E,t) \neq P(S_2 \times S_2, t)\cdot P(S_3, t) = P(S_2 \times S_2 \times S_3, t).$$

<u>Problem</u> (Koszul). If the structure group is a connected Lie group does it follow that if π^* is injective then i^* is surjective? In general the answer is unknown. We shall later prove a result of Kudo that the implication holds in the case of principal bundles.

15. <u>Pairing of the spectral sequence of a principal bundle with the homology of the structural group.</u>

Consider a principal bundle E with structural group X. Correspondingly we have a commutative diagram

$$\begin{array}{ccc} E \times X & \xrightarrow{\Phi} & E \\ \pi \downarrow \quad \downarrow & & \downarrow \pi \\ B \times p & \xrightarrow{\bar{\Phi}} & B \end{array}$$

in which Φ defines the operations of X on E, p is a fixed point of B, and $\bar{\Phi}$ is defined by $(b,p) \to b$.

At the same time we wish to consider a more general situation. Let E be a topological space on which an H-space X operates freely, by which we mean that only the identity operation has fixed points, and assume that the projection π of E onto its quotient space B (relative to the equivalence relation defined by the operations of X) is a fibre map (say in Serre's sense). Thus we get a bundle satisfying conditions (a) and (b) for a principal bundle. Apart from principal bundles, the standard case is where E is the space of paths ending at a fixed point of a given topological space and X is the space of loops at the point with composition of paths defining the operations. Then the fibres are of the same homotopy type as X. This case is studied in [3].

We shall call this a <u>quasi-principal bundle</u>.

In what follows we assume always a coefficient field K which we shall omit writing.

<u>Lemma 15.1</u>. Let (E_i,B_i,F_i,π_i), $(i=1,2)$, be two fiberings with $\tilde{H}^*(F_i)$ simple, and consider the fibering

$$(E,B,F,\pi) = (E_1 \times E_2, B_1 \times B_2, F_1 \times F_2, \pi_1 \times \pi_2).$$

Denote the respective (cohomology) spectral sequences by $({}_iE_r)$, $(i=1,2)$, and (E_r). Then there is an isomorphism $E_r = {}_1E_r \otimes {}_2E_r$ which is compatible with all structures.

<u>Proof</u>. Consider the commutative diagram

$$\begin{array}{ccccc} E_1 & \xrightarrow{\alpha_1} & E_1 \times E_2 & \xrightarrow{\beta_1} & E_1 \\ \pi_1 \downarrow & & \downarrow \pi_1 \times \pi_2 & & \downarrow \pi_1 \\ B_1 & \xrightarrow{\bar{\alpha}_1} & B_1 \times B_2 & \xrightarrow{\bar{\beta}_2} & B_1 \end{array}$$

where α_1 and β_1 are given by $e_1 \to (e_1, e_2^0)$ and $(e_1, e_2) \to e_1$, respectively, and $\bar{\alpha}_1$, and $\bar{\beta}_1$ are the induced maps. As in (13.7) α_1, β_1 induce maps on the spectral sequences,

$$\beta_1^* : ({}_1E_r) \to (E_r), \quad \alpha_1^* : (E_r) \to ({}_1E_r)$$

such that $\alpha_1^* \beta_1^*$ is the identity map. Then it follows that ${}_1E_r$ is isomorphic under β_1^* to a subalgebra of E_r. Similarly ${}_2E_r$ is isomorphic to a subalgebra of E_r. We define

$$\beta^* = \beta_1^* \otimes \beta_2^* : {}_2E_r \otimes {}_1E_r \to E_r$$

by $\beta^*(a \otimes b) = \beta_2^*(a) \cdot \beta_1^*(b)$. This defines β^* as a multiplicative homomorphism which is compatible with d_r and the total differential on ${}_2E_r \otimes {}_1E_r$. For $r = 2$ we have

$$E_r \approx H^*(B_1 \times B_2) \otimes H^*(F_1 \times F_2), \text{ and from section 13}$$

and the details of the Künneth rule we see that

$$E_2 \approx H^*(B_1) \otimes H^*(B_2) \otimes H^*(F_1) \otimes H^*(F_2) \approx {}_1E_2 \otimes {}_2E_2.$$

It follows trivially that $E_r \approx {}_1E_r \otimes {}_2E_r$ $(2 \leq r \leq \infty)$ and the lemma is proved.

A similar lemma can be proved for the homology spectral sequences.

We apply the lemma (15.1) to the two fiberings which appear in the initial diagram of this section. The second fibering $X \to p$ is of course trivial; we have

$${}_2E_2^{p,q} = \begin{cases} 0 & \text{if } p > 0 \\ H^q(X), & d_r \equiv 0. \end{cases}$$

Combining the isomorphism of the lemma with the induced map ϕ^* gives a map, which we also denote by ϕ^*,

$$\phi^* : E_r \to E_r \otimes H^*(X);$$

more precisely,

$$\phi^* : E_r^{p,q} \to \sum_{i>0} E_r^{p,q-i} \otimes H^i(X).$$

In a similar fashion using the analogous lemma for homology and the induced map ϕ^* we obtain a map

$$\phi_* : E^r \otimes H_*(X) \to E^r$$

which has the following properties:

(a) ϕ_* is a homomorphism $E_{p,q}^r \otimes H_i(X) \to E_{p,q+i}^r$. If $a \in E_{p,q}^r$ and $u \in H_i(X)$ we denote $\phi_*(a \otimes u)$ by $a \vee u$.

(b) If we identify $E^2_{p,q} = H_p(B) \otimes H_q(X)$ then

$$(b \otimes u) \vee v = b \otimes (u \vee v);\ b \in H_p(B),\ u \in H_q(X),\ v \in H_i(X),$$

where the second $\vee$ denotes the Pontrjagin product in $H_*(X)$.

(c) $$(a \vee u) \vee v = a \vee (u \vee v);\ a \in E^r_{p,q},\ \ u,v \in H_*(X),$$

where the second $\vee$ denotes the Pontrjagin product in $H_*(X)$.

(d) ϕ_* commutes with d_r and k^r_{r+1}.

(e) If J_i is the ith level in the filtration of $H_*(E)$ then $J_i \vee H_*(X) \subset J_i$, and moreover the induced operation of $H_*(X)$ on $GrH_*(E)$ is given by the induced homomorphism $E^\infty \otimes H_*(X) \to E^\infty$.

The map ϕ^* is the analogue of the Pontrjagin product which we defined in section ' Continuing the analogy we now define a "cap" product $E_r \otimes H_*(X) \to E_r$ as follows. Let $a \in E_r$; then ϕ^* (a) is in $E_r \otimes H^*(X)$, and we can write

$$\phi^*(a) = \sum a_i \otimes x_i,\ a_i \in E_r^{p,q-i},\ x_i \in H^i(X).$$

Set $a \vee u = \sum a_i\langle u,x_i\rangle$; then the following properties hold:

(a) $a \vee u$ pairs $E_r^{p,q}$, $H_i(X)$ to $E_r^{p,q-i}$.

(b) If we identify $E_2^{p,q} = H^p(B) \otimes H^q(X)$ then

$$(b \otimes x) \vee u = b \otimes (x \vee u),\ b \in H^p(B),\ x \in H^q(X),\ u \in H_i(X),$$

where the second $\vee$ denotes the Pontrjagin "cap" product in X.

(c) $$(a \vee u) \vee v = a\ (u \vee v),\ a \in E_r^{p,q},\ \ u,v \in H_*(X),$$

where the second $\vee$ denotes the Pontrjagin product in X.

(d) $\vee$ commutes with d_r and k^r_{r+1}.

(e) If J_i is the ith level in the filtration of $H^*(E)$ then $J_i \vee H\ (X) \subset J_i$, and the induced operation on $GrH^*(E)$ coincides with the induced map $E_\infty \vee H_*(X) \to E_\infty$.

(f) If u is a homogeneous element of $H_*(X)$ and is orthogonal to the decomposable elements in $H^*(X)$ and a_1, a_2 are homogeneous elements of E_r then

$$(a_1 a_2) \vee u = a_1 \cdot (a_2 \vee u) + (-1)^{d^\circ u \cdot d^{a_2}}(a_1 \vee u) \cdot a_2.$$

The verifications of properties (a,b,c,d) for the homology pairing and (a,b,c,d,f) for the cohomology pairing are similar to those of section 7 making use also of lemma (15.1) and its homology analog. The proof of (e) requires explicit knowledge of the filtration, see [3].

<u>Theorem 15.2</u>. (Kudo [7]). Let (E,B,X,π) be a quasi-principal bundle. If π^* is injective then i^* is surjective (and $E_2 = E_\infty$).

<u>Proof</u>. Since the fibres are connected $H^*(X)$ is a simple system. Thus

$$E_2^{p,q} = H^p(B) \otimes H^q(X).$$

Assume that i^* is not surjective; then for some first index $s, d_s \neq 0$. We have

$$E_s^{p,q} = E_2^{p,q} = H^p(B) \otimes H^q(X).$$

Since $d_s \neq 0$ there is an integer q and an element $x \in H^q(X)$ such that $d_s(1 \otimes x) \neq 0$. We can write

$$d_s(1 \otimes x) = \sum b_i \otimes x_i,$$

where the b_i are non-zero elements of $H^s(B)$ and the x_i are linearly independent elements of $H^{q-s+1}(X)$. Let $u_1 \in H_*(X)$ be dual to x_1; then

$$d_s(x \vee u_1) = (d_s x) \vee u_1 = (\sum b_i \otimes x_i) \vee u_1 = \sum b_i \otimes (x_i \vee u_1),$$
$$= \sum b_i \otimes 1 \langle x_i, u_1 \rangle = b_1 \otimes 1.$$

Thus $b_1 \otimes 1$ is a coboundary so that $k_{s+1}^s b_1 = 0$. It follows that the projection of b_1 in E_∞ is zero and $E_2^{s,0} \to E_\infty^{s,0}$ is not injective. But then π^* is not injective in dimension s which contradicts the hypothesis. Thus i^* must be surjective.

Theorem 15.3. Let (E,B,X,π) be a quasi-principal bundle with $H^q(E) = 0$ for $q > 0$. Then every transgressive element $x \in H^q(X)$ is also primitive.

Proof. We have $E_2^{0,q} = H^q(X)$. An element $x \in H^q(X)$ is transgressive if $d_r k_r^2 x = 0$ for $r \leq q$. On the other hand x is non-primitive means there is an $u \in H_*(X)$, $d^\circ u < d^\circ x$, such that $x \vee u \neq 0$ where $\vee$ denotes the Pontrjagin "cap" product. We shall show these two conditions are incompatible under the hypothesis of the theorem. We have

$$d_r k_r^2(x \vee u) = (d_r k_r^2 x) \vee u = 0 \quad \text{for } r \leq q.$$

Therefore

$$d_r k_r^2(x \vee u) = 0 \quad \text{for } r \leq d^\circ(x \vee u) + 1;$$

but then by a general property of the spectral sequence

$$d_r k_r^2(x \vee u) = 0 \quad \text{for } r > d^\circ(x \vee u) + 1$$

so that $x \vee u$ is a permanent cocycle. If we set $d^\circ(x \vee u) = m$ we have $E_\infty^{0,m} \neq 0$, and it follows that $H^m(E) \neq 0$. Since $m > 0$ this contradicts the hypothesis. Thus if u is transgressive it must be primitive.

Remark. One can develop similar considerations in other cases where a group operates, for example in the following case first considered by J. Leray [8]. Let (E,B,F,π) be a bundle (or fibre space) and X be an H-space operating on E and B, the operations commuting with π. Then we have a commutative diagram

$$\begin{array}{ccc} E \times X & \longrightarrow & E \\ \downarrow \pi \quad \downarrow id & & \downarrow \pi \\ B \times X & \longrightarrow & B \end{array}$$

The spectral sequence of $X \xrightarrow{id} X$ is of course given by ${}_2E_2^{p,q} = 0$ $(q > 0)$, ${}_2E_r^{p,o} = H^p(X)$; $d_r \equiv 0$, and using lemma 15.1 we get homomorphisms

$$\phi^* : E_r^{p,q} \to \sum E_r^{p-i,q} \otimes H^i(X),$$

and in homology,

$$\phi_* : E^r_{p,q} \otimes H_s(X) \to E^r_{p+s,q}.$$

These again give rise to Pontrjagin products and "cap" products, which now alter the base-degree but otherwise have properties similar to those above. For instance, in cohomology we have

(a') $a \vee u$ pairs $E_r^{p,q}$, $H_i(X)$ to $E_r^{p-i,q}$,

(b') $(b \otimes x) \vee u = (b \vee u) \otimes x$,

and (c) to (f) as before.

A further generalization would be the case where there is a homomorphism f of X onto another H space Y operating on Y, such that

$$\begin{array}{ccc} E \times X & \longrightarrow & E \\ \downarrow \pi \quad \downarrow f & & \downarrow \pi \\ B \times Y & \longrightarrow & B \end{array}$$

is commutative. We conclude the chapter with the following digression.

<u>Theorem 15.4</u>. If Ω_n denotes the space of loops on the sphere S_n, $n \geq 2$, then

$$H_*(\Omega_n, Z) = Z[u], \; d^o u = n-1,$$

the product being the Pontrjagin product.

<u>Proof</u>. Let E be the space of paths on S_n starting at a fixed point; then E is acyclic and the map $\pi : E \to S_n$ defined by projecting each path to its end point is a fibre map in the sense of Serre with Ω_n as fibre. In the corresponding (homology) spectral sequence we have

$$E^2 \approx H_*(S_n, Z) \otimes H_*(\Omega_n, Z),$$

$$E^\infty_{o,o} = Z, \; E^\infty_{p,q} = 0 \quad \text{for } p+q>0.$$

One shows readily that d_r vanishes except for

$$d_n : E^n_{n,k(n-1)} \to E^n_{o,(k+1)(n-1)}, k = o,1,2,\ldots,$$

which are isomorphisms. It follows that

$$H_i(\Omega_n, Z) = \begin{cases} Z & \text{for } i = k(n-1), \\ 0 & \text{otherwise.} \end{cases}$$

For the details see [12]. Now let x be a generator for $H_n(S_n, Z)$. We can choose $e_k \in H_{k(n-1)}(\Omega_n, Z)$ such that

$$d_n(x \otimes 1) = e_1, \quad d_n(x \otimes e_{k-1}) = e_k.$$

We shall prove by induction on k that $e_k = e_1^k$ (= $e_1 \vee \cdots \vee e_1$, k times).

Assuming the result up to k then

$$e_{k+1} = d_n(x \otimes e_1^k) = d_n((x \otimes e_1^{k-1}) \vee e_1) = (d_n(x \otimes e_1^{k-1})) \vee e_1,$$
$$= d_n(x \otimes e_{k-1}) \vee e_1 = e_k \vee e_1 = e_1^{k+1},$$

and the induction is completed.

Applying the converse of (5.3) $H^*(\Omega_n, Z)$ is a twisted polynomial ring. A result in [3] generalizes (15.4) to the case of finitely many copies of S_n joined at a common point.

BIBLIOGRAPHY

[1] A. Borel, Séminaire de Topologie algébrique de l'E.P.F., Zurich (1951).

[2] A. Borel, Sur la cohomologie des espaces fibrés principaux et des espaces homogènes de groupes de Lie compacts, Ann. of Math. 57 (1953), 115-2o7.

[3] R. Bott and H. Samelson, On the Pontrjagin product in spaces of paths, Comment. Math. Helv. 27 (1953), 32o-337.

[4] H. Cartan, Séminaire de Topologie algébrique de l'E.N.S. II, Paris (1949-5o).

[5] H. Cartan, idem III, (195o-51).

[6] S. Eilenberg, idem III, (195o-51).

[7] T. Kudo, Homological structure of fibre bundles, Jour. Inst. of Polyt. Osaka City Univ. 2 (1952), 1o1-14o.

[8] J. Leray, Espaces où opère un groupe de Lie compact connexe, Comptes rendus de l'Académie des Sciences, Paris, t. 228 (1949), pp. 1545-47.
- - -, Applications continues commutant avec les éléments d'un groupe de Lie, ibid., pp. 1784-86.

[9] J. Leray, L'homologie d'un espace fibré dont la fibre est connexe, Jour.Math. Pures Appl. IXs. 29 (195o), 169-213.

[1o] S.D. Liao, On the theory of obstructions of fibre bundles, Ann. of Math. (2) 6o (1954), 146-191.

[11] S. Mac Lane, Spectral sequences and homotopy, U. of Chicago (mimeographed note).

[12] J.P. Serre, Homologie singulière des espaces fibrés. Applications, Ann. of Math. 54 (1951), 425-5o5.

[13] N. Steenrod, The topology of fibre bundles, Princeton (1951).

CHAPTER III

UNIVERSAL BUNDLES AND CLASSIFYING SPACES

16. Universal bundles and classifying spaces.

Let G be a compact Lie group. An n-universal bundle for G is a principal bundle E(n,G) whose structural group is G and whose first n homotopy groups vanish. The base space B(n,G) is said to be an n-classifying classifying space for G, as is suggested by the following classification theorem.

Theorem 16.1. Let B be an n-dimensional polyhedron. Then the equivalence classes of principal bundles with structural group G and base space B are in 1 - 1 correspondence with the homotopy classes of maps B → B (n,G), the bundle corresponding to a given map being the induced bundle.

For the proof see section 19 [6], (and [5] in the case where B is a locally compact space). The existence of n-universal bundles (in fact, compact differentiable bundles) for any n > o and any compact Lie group G is well known (see [6], [5]). Briefly, if G = O(m) we take for E(n,O(m)) the Stiefel manifold $W_{m+n+1,m}$ = O(m+n+1) / O(n+1); then B(n,O(m)) is the Grassmann manifold $G_{m+n+1,m}$ = O(m+n+1) / O(n+1) × O(m). Since for any closed subgroup U ⊂ G an n-universal bundle E(n,G) is also an n-universal bundle E(n,U), it follows that there exist n-universal bundles for any closed subgroup of O(m). Making use of the Peter-Weyl representation theorem we then have the existence of n-universal bundles for any compact Lie group.

Proposition 16.2. Let E_1,E_2 be n-universal bundles for G with B_1,B_2 the corresponding n-classifying spaces. Then for any coefficient ring A there is an isomorphism $H^i(B_1,A) = H^i(B_2,A)$ compatible with the product up to n,(i.e., $i \leq n$).

Consider the commutative diagram

$$
\begin{array}{ccccc}
E_1 & \xleftarrow{\ f_1\ } & E_1 \times E_2 & \xrightarrow{\ f_2\ } & E_2 \\
\Big\downarrow{\scriptstyle \pi_1} & & \Big\downarrow{\scriptstyle \pi} & & \Big\downarrow{\scriptstyle \pi_2} \\
B_1 & \xleftarrow{\ \bar{f}_1\ } & (E_1,E_2)_G & \xrightarrow{\ \bar{f}_2\ } & B_2
\end{array}
$$

where f_1,f_2 are the natural projections and $\bar{f}_1,\bar{f}_2$ the induced maps. Then $\bar{f}_1,\bar{f}_2$ are fibre maps with E_2,E_1 as fibres respectively. By the following lemma $\bar{f}_1^*$ and $\bar{f}_2^*$ are isomorphisms up to dimension n. This proves 16.2. (Note that the isomorphisms are canonical in the sense of transitivity.)

Lemma. Let (E,B,F,π) be a fibre bundle with $H^j(F,A) = 0$ for $0 < j \leq n$. Then $\pi^*: H^i(B,A) \to H^i(E,A)$ is an isomorphism up to n.

Proof. We have

$$E_2^{i,j} \approx \begin{cases} H^i(B,A) & \text{if } j = 0 \\ 0 & \text{if } 0 < j \leq n. \end{cases}$$

Thus an element of E_2 of total degree $i+j \leq n$ must be zero if $j \neq 0$, and it follows that $H^i(E,A) = E_\infty^{i,o}$. Moreover every element of $E_2^{i,o}$ is a d_2-cycle and none is a d_2-boundary. Thus for $i \leq n$, π^* is given by

$$H^i(B,A) \approx E_2^{i,o} = \ldots = E_\infty^{i,o} = H^i(E,A).$$

Proposition 16.2 allows us to define a graded algebra $H^*(B_G, A)$ which for each n is isomorphic up to n with $H^*(B(n,G),A)$. A more concrete way of defining $H^*(B_G,A)$ and to "pass to the limit $n \to \infty$" is to introduce a universal bundle E_G for G, (that is, an n-universal bundle for all n.) The bundle E_G may be defined as the inductive limit of a sequence of n_i-universal bundles E_i $(n_1 < n_2 < \cdots)$ together with injections $f_i : E_i \to E_{i+1}$. The existence of such sequences follows from the fact that we may take as E_i Stiefel manifolds. Then E_G is a principal bundle which is acyclic in all dimensions, and its base space B_G, determined up to homotopy type, is a classifying space for G for all n.

Let E be a principal bundle for G with base space B, then by the classification theorem it is induced by a map $\phi : B \to B_G$. The map $\phi^* : H^*(B_G,A) \to H^*(B,A)$ is thus uniquely determined by the bundle structure of E and is called the characteristic map. Its image is called the characteristic ring of E. Of course ϕ also defines a homomorphism of the spectral sequence of (E_G,B_G,G) into that of (E,B,G).

Using the diagram in 16.2 and property 13.8 we see that when G is connected there is a canonical isomorphism up to total degree n-dim G between the spectral sequences of two n-universal bundles for G. We may therefore define a universal spectral sequence for G as a spectral sequence $(E_r)_G$, $r \geq 2$, which for each n is isomorphic up to n-dim G with the spectral sequence of $(E(n,G),B(n,G),G)$. Clearly it is nothing other than the spectral sequence of a universal bundle (E_G,B_G,G) taken either in singular or Alexander-Spanier cohomology. A map $\phi : B \to B_G$ which induces a given bundle (E,B,G) then defines a homomorphism of $(E_r)_G$ into the spectral sequence of the bundle.

<u>Proposition 16.3</u>. Let (E,B,F,π) be a fibering with group a compact Lie group G. Suppose E and F are principal bundles for G such that its operations on E and F commute with the admissible maps between the fibres F_b and the standard fibre F of the original bundle. Then we can define a new fibering $(E/G,B,\ F/G,\bar{\pi})$ with $\bar{\pi}$ defined naturally. Let $\bar{i} : F/G \to E/G$ be the natural map. We assert that the image of

$$\bar{i}^*: H^*(E/G,A) \to H^*(F/G,A)$$

contains the characteristic ring of $(F,F/G,G)$.

<u>Proof</u>. Consider the diagram

$$\begin{array}{ccccc} F & \xrightarrow{i} & E & \longrightarrow & E_G \\ \downarrow & & \downarrow & & \downarrow \\ F/G & \xrightarrow{\bar{i}} & E/G & \xrightarrow{\phi} & B_G \end{array}$$

where ϕ induces the principal bundle structure of E. Since the principal bundle $(F,F/G,G)$ is equivalent to the induced bundle of the map $\phi\bar{i}$ the assertion follows at once from the classification theorem.

<u>Corollary 16.4</u>. If $A = K$, and $H^*(F/G,K)$ is the characteristic ring of $(F,F/G,G)$, then F/G is totally non-homologous to 0 in any bundle $(E/B,B,F/G,\bar{\pi})$.

Given a principal bundle E with group G and a topological space F on which G acts (on the right) we have earlier defined an associated bundle $((E,F)_G,B,F,\bar{\pi})$ with base space $B = E/G$. Now given a group G and a topological space F on which G operates (on the right) we define a <u>universal associated bundle</u> for the pair (G,F) as a bundle $((E_G,F)_G,B_G,F,\bar{\pi})$ where E_G is a universal bundle for G. Note however that $(E_G,F)_G$ is not acyclic. Corresponding to any bundle (E,B,F,π) with group G we can write a diagram

$$\begin{array}{ccc} E & \xrightarrow{\tilde{\phi}} & (E_G,F)_G \\ \pi\downarrow & & \downarrow\bar{\pi} \\ B & \xrightarrow{\phi} & B_G \end{array} .$$

The induced maps $\tilde{\phi}^*$ and ϕ^* are <u>characteristic maps</u> and the corresponding rings <u>characteristic rings</u> for the bundle. The spectral sequence of $(E_G,F)_G$ is the <u>universal spectral sequence</u> for (G,F); it is mapped homomorphically by $\tilde{\phi}^*$ into that of (E,B,F). Note that if $F = G/U$ then $(E,F)_G$ may be identified with E/U.

17. $\rho(U,G)$: 3 fiberings involving classifying spaces.

If U is a closed subgroup of G then E_G is also an E_U. The inclusion $U \subset G$ then gives rise to the fibre map $\rho(U,G) : B_U \to B_G$ of the natural fibering $(B_U, B_G, G/U, \rho(U,G))$. We can generalize $\rho(U,G)$ as follows. Let $f : U \to G$ be a homomorphism and let U operate on G by $T_u(g) = f(u) \cdot g$. We can then form the bundle $((E_U,G)_U, B_U, G)$. Since the operations of T_u commute with G operating on itself by right translations, it is easily seen that we obtain a principal bundle with group G. It is induced by a map of B_U into B_G, denoted by $\rho(f)$, which generalizes $\rho(U,G)$.

Beginning with a fibering (E,B,F,π) we may derive a new fibering (E',B',F',π') in three special cases listed in the following table:

Case	E	B	F	π	E'	B'	F'	π'	i'
I. Principal bundle	X	Y	U	π	$\approx Y$[1]	B_U	X	χ	π
II. U closed subgroup of G	G	G/U	U	π	B_U	B_G	G/U	$\rho(U,G)$	χ
III. U closed and invariant	G	G/U	U	π	$\approx B_G$[1]	$B_{G/U}$	B_U	$\rho(\pi)$	$\rho(U,G)$

(χ is the classifying map of the given bundle.)

We remark that U is a compact Lie group which is not necessarily connected and may possible be finite. In the latter case $H^*(B_U)$ is the cohomology of U in the sense of group theory. We have already considered II above. The derived fibering in III is useful in connection with group extensions.

Case I. We consider two fiberings $((E_U,X)_U, B_U, X)$ and $((E_U,X)_U, Y, E_U)$. In the latter the fiber E_U is acyclic, so $(E_U,X)_U \approx Y$. Thus the first fibering corresponds to the derived fibering in I. To show that the fibre map induces the classifying map of the original bundle we consider the diagram

$$\begin{array}{ccccc} E_U & \xleftarrow{g} & E_U \times X & \xrightarrow{f} & X \\ \downarrow & & \downarrow & & \downarrow \\ B_U & \xleftarrow{\bar{g}} & (E_U \times X)_U & \xrightarrow{\bar{f}} & Y \end{array}$$

where f,g are the natural projections. The map $\bar{f}^*$ is an isomorphism. We have to show that $(\bar{f}^*)^{-1}\bar{g}^*$ is the characteristic map ϕ^* of (X,Y,U). Let $\tilde{\phi} : X \to E_U$ correspond to ϕ, and define $\Psi : X \to E_U \times X$ by $x \to (\tilde{\phi}(x), x)$. Then $g\Psi = \tilde{\phi}$ and $f\Psi =$ identity. It follows that $(\bar{f}^*)^{-1}\bar{g}^* = \phi^*$. It is also easy to see that $i' = \pi$.

(1) $\approx X$ means a space whose homology or homotopy groups are canonically isomorphic to those of X.

Case III. Consider the fibering ($E_G/U = B_U, B_G, G/U$) of II. Since U is invariant this is a principal bundle with fibre G/U. Then $\bar{E} = (E_G/U, E_{G/U})_{G/U}$ has two fiberings $(\bar{E}, B_G, E_{G/U})$ and $(\bar{E}, B_{G/U}, B_U)$; the fibre of the first being acyclic we see that $\bar{E} \approx B_G$ from which it follows that the second fibering gives the derived fibering in III. We leave the details to the reader.

18. Some results on universal spectral sequences.

We have previously defined the universal spectral sequence $(E_r)_G$ for a compact connected Lie group G and observed that if (E,B,G) is induced by $\phi : B \to B_G$ then ϕ defines a homomorphism of $(E_r)_G$ into the spectral sequence of (E,B,G). It follows that properties of the former have a "universal" character in the sense that they are valid in any principal bundle. For instance if $x \in H^*(G,A)$ is transgressive in E_G then it will be transgressive in any principal bundle for G. We call such elements universally transgressive. The study of $(E_r)_G$ is an important problem which has not yet been fully investigated. We list below some results obtained so far. A part of these results will be proved in the next section.

Theorem 18.1. Let $H^*(G,K_p) = \Lambda(x_1,\ldots,x_m)$, $d^o x_i$ odd, then

(a) $H^*(G,K_p) = \Lambda(x_1',\ldots,x'_m)$ $d^o x_i' = d^o x_i$,

where the x_i' are universally transgressive;

(b) $H^*(B_G,K_p) = K_p[y_1,\ldots,y_m]$, $d^o y_i = 1 + d^o x_i$,

the y_i being images of the x_i' by transgression.

A similar statement is valid for integer coefficients. As a converse to (b) we have the following:

Theorem 18.2. If $H^*(B_G,K_p) = K_p[y_1,\ldots,y_m]$, $d^o y_i$ even, then

$H^*(G,K_p) = \Lambda(x_1,\ldots,x_m)$, $d^o x_i = d^o y_i - 1$,

where the x_i are universally transgressive elements whose images are the y_i, and similarly with Z in place of K_p.

Theorem 18.3. If $H^*(G,K_2) = \Delta(x_1,\ldots,x_m)$ where the x_i are universally transgressive then $H^*(B_G,K_2) = K_2[y_1,\ldots,y_m]$ where the y_i are images of the x_i by transgression, and conversely.

For the proofs we refer the reader to [1] and [2]. As a consequence of (18.1) we have the following:

Proposition 18.4. If $H^*(G,K_p) = \Lambda(x_1,\ldots,x_m)$ $d^o x_i$ odd, then G and B_G have no p-torsion.

Proof. From (18.1b) we see that $H^*(B_G,K_p)$ has only even degrees; hence it has no p-torsion, and $P_p(B_G,t) = P_o(B_G,T)$. By Hopf's theorem the assumption of 18.1 always holds in characteristic zero; thus we see that $H^*(B_G,K_p)$ and $H^*(B_G,K_o)$ are both rings of polynomials over elements which, because of the equality of the Poincaré polynomials, must have the same degrees. Then (18.2) gives $P_p(G,t) = P_o(G,t)$ so G has no p-torsion.

A comparison of Poincaré polynomials and application of (18.4) then gives the following:

Proposition 18.5. (a) If G has no torsion then B_G has no torsion.

(b) G has no p-torsion if and only if

$$H^*(G, K_p) = \Lambda(x_1,\ldots,x_m),\ d^o x_i \text{ odd.}$$

19. Proof of one theorem on universal spectral sequences.

We shall prove the following theorem:

Theorem 19.1. If $H^*(G,K_p)$ has a simple system of universally transgressive elements $x_1,\ldots,x_m$ then

$$H^*(B_G,K_p) = K_p[y_1,\ldots,y_m]$$

where y_i is an image of x_i by transgression.

Notice that for $p \neq 2$ our assumption implies that $H^*(G,K_p) = \Lambda(x_1,\ldots,x_m)$, $d^o x_i$ odd, (proposition 3.5), so that the y_i have even degrees and the statement is in agreement with anti-commutativity. The theorem includes 18.3, and for $p \neq 2$ is weaker than 18.1 since it gives (b) if we assume (a). Although a more "conceptual" proof can be given, we follow here the method of [1] since it is also used to prove theorem 18.1.

We begin with some algebraic preliminaries. Let B be an algebra over a field K, graded by B^i $(i \geq 0)$, anti-commutative, and with B^o spanned by the unit element 1. Let $a_1,\ldots,a_m$ be homogeneous elements of B; we shall say they are ann-free (annihilator-free) up to k if the annihilator of the projection of a_i in $B/(a_1,\ldots,a_{i-1})$ is zero up to degree $k-d^o a_i$ for all $i = 1,2,\ldots,m$, (for $i = 1$ we mean the annihilator of a_1 in B vanishes up to degree $k - d^o a_1$). If $k \geq d^o a_i$, $(1 \leq i \leq m)$, this implies linear independence. If the a_i are ann-free up to k for all k we simply say they are ann-free.

One shows readily by induction that if the a_i are ann-free they are algebraically independent. It will be apparent by 19.2 (and could also be seen directly) that if the a_i are ann-free up to k, $(k \geq d^o a_i)$, then so are the elements of any basis of the vector space Q spanned by the a_i. Thus any permutation of the a_i are ann-free up to k. We may also speak of Q as being ann-free. In what follows we will have to consider not only an algebra $\Delta(x_1,\dots,x_m)$ but also certain of its subspaces spanned by some of the x_i and their products $x_{i_1}\cdots x_{i_k}$, $(i_1<\dots<i_k)$. Such subspaces retain a part of the multiplicative structure of $\Delta(x_1,\dots,x_m)$, (for example, if x_i and x_j are in the subspace and $i<j$, their product x_ix_j is also in the subspace), but in general they are not subalgebras. In order to give them a more precise formal description we shall denote by $\Delta^{\dagger}(x_1,\dots,x_m)$ a graded vector space which (a) is a module over the exterior algebra $\Lambda(x_1,\dots,x_m)$, (b) has a basis composed of an element 1 of degree 0 and its transforms $x_{i_1}\cdots x_{i_k}\cdot(1)$, $(i_1<\dots<i_k)$; the latter elements will also be written $x_{i_1}\cdots x_{i_k}$. The grading in $\Delta^{\dagger}(x_1,\dots,x_m)$ is given by

$$d^{\circ}(x_{i_1}\cdots x_{i_k}\cdot(1)) = \sum_{j=1}^{k} d^{\circ}x_{i_j}.$$

There is an obvious isomorphism

$$\Delta^{\dagger}(x_1,\dots,x_k)\otimes\Delta^{\dagger}(x_{k+1},\dots,x_m)\approx\Delta^{\dagger}(x_1,\dots,x_m)$$

which maps $1\otimes 1$ to 1, $x_i\otimes 1$ to x_i, $1\otimes x_j$ to x_j, $x_i\otimes x_j$ to x_ix_j, etc. Evidently every $\Delta(x_1,\dots,x_m)$ may be regarded as a $\Delta^{\dagger}(x_1,\dots,x_m)$. Of course we could identify $\Delta^{\dagger}(x_1,\dots,x_m)$ with $\Lambda(x_1,\dots,x_m)$, but for the reasons mentioned above we wish to avoid it and rather regard a product as an external law.

Let $J = B\otimes\Delta^{\dagger}(x_1,\dots,x_m)$. Then J is a vector space graded in a natural way, and we denote by DB, DF, and D respectively, the base degree (d° in B), the fibre degree (d° in $\Delta^{\dagger}(x_1,\dots,x_m)$), and the total degree (D = DB + DF). In addition J is graded by subspaces $B\otimes^{(e)}\Delta^{\dagger}(x_1,\dots,x_m)$, spanned by the generators $b\otimes x_{i_1}\cdots x_{i_e}$, and call e the <u>exterior</u> degree. We may consider J as a module over $B\otimes\Lambda(x_1,\dots,x_m)$ in an obvious way. In particular, if $h = b\otimes x_{i_1}\cdots x_{i_e}$ we have

$$(a\otimes 1)h = ab\otimes x_{i_1}\cdots x_{i_e},$$
$$(1\otimes x_1)h = b\otimes x_1x_{i_1}\cdots x_{i_e}\qquad \text{if } i_1\geq 2).$$

We shall hereafter assume that $d^{\circ}x_i = d^{\circ}a_i-1$ where $a_1,\dots,a_m$ are elements of degree s in the center of B. If $p\neq 2$ this implies that s is even and $d^{\circ}x_i$ odd.

We define a differential d on J by

$$dh = d(b\otimes x_{i_1}\cdots x_{i_e}) = \sum(-1)^{j-1}ba_{i_j}\otimes x_{i_1}\cdots\hat{x}_{i_j}\cdots x_{i_e},$$

($\hat{}$ indicates ommission). We have clearly

$$d((a\otimes 1)h) = (a\otimes 1)dh,$$
$$d((1\otimes x_1)h) = (a_1\otimes 1)h-(1\otimes x_1)dh\qquad(\text{for } i_1\geq 2).$$

Then d is homogeneous with respect to all gradings of J, increasing DB by s, decreasing DF by s-1, increasing D by 1, and decreasing the exterior degree by 1. The homology of J with respect to d (coefficients in K) is then naturally graded by $H(J) = \sum_{p,q,e} H_e^{p,q}$; in particular we note $H_o(J) = B/(a_1,\dots,a_m)$.

Lemma 19.2. The following conditions are equivalent:

(1) $a_1,\dots,a_m$ are ann-free up to k,

(2) $H_e^{p,q} = 0$ for $p \le k - s$, $(e \ge 1)$,

(3) $H_1^{p,q} = 0$ for $p \le k - s$.

Proof. For m = 1 the lemma is trivial. Assume the lemma true for m-1. Let

$$J_1 = B \otimes \Delta^{\dagger}(x_2,\dots,x_m),$$
$$J_1' = B/(a_1) \otimes \Delta^{\dagger}(x_2,\dots,x_m),$$

and let $\lambda : J_1 \to J'_1$ be the natural map. Evidently J_1 is a subspace of J, in fact a $(B \otimes \Lambda(x_2,\dots,x_m))$ submodule stable under d. The kernel of λ is clearly $(a_1) \otimes \Delta^{\dagger}(x_2,\dots,x_m)$. We define a differential d' on J_1' by

$$d(b' \otimes x_{i_1}\cdots x_{i_k}) = \sum (-1)^{j-1} b' a'_{i_j} \otimes x_{i_1}\cdots \hat{x}_{i_j}\cdots x_{i_k},$$

$(b' \in B',\ a'_{i_j} = \lambda(a_{i_j}))$, and then $d'\lambda = \lambda d$. By our inductive assumption the lemma holds for J_1', and it remains to infer it for J.

(1)$\Longrightarrow$(2) : The $a_1,\dots,a_m$ being ann-free up to k, the same is then true for $a_2',\dots,a_m'$ and so (2) holds for J_1'. Let $h \in J$, $DBh \le k-s$, $DFh > 0$, be a cocycle. We have to show it is a coboundary. Write

$$h = (1 \otimes x_1)h_1 + h_2;\ h_1,h_2 \in J_1,$$

with $DBh_1 = DBh_2 = DBh$ and $DFh_2 = DFh > 0$. Then

$$dh = (a_1 \otimes 1)h_1 + (-1)^{s-1}(1 \otimes x_1)dh_1 + dh_2,$$

the first and last terms on the right side being in J_1. Since dh = 0 we have

$$(a_1 \otimes 1)h_1 + dh_2 = 0,\quad (1 \otimes x_1)dh_1 = 0.$$

Projecting into J_1' we see that $d'h_2' = 0$. Therefore in J_1' we have $h_2' = d'h_3'$ for some $h_3' \in J_1'$ with $DBh_3' = k-2s$: hence

$$h_2 = dh_4 + (a_1 \otimes 1)h_4,\ DBh_3 = DBh_4 = k-2s,$$

with $h_3,h_4 \in J_1$. Substituting,

$$h = (1 \otimes x_1)h_1 + dh_3 + (a_1 \otimes 1)h_4,$$
$$dh = (a_1 \otimes 1)(h_1 + dh_4) + (-1)^{s-1}(1 \otimes x_1)dh_1 = 0.$$

Thus $(a_1 \otimes 1)(h_1 + dh_4) = 0$. The annihilator of $(a_1 \otimes 1)$ is identical with (the annihilator of a_1 in B) $\otimes$ 1 which vanishes up to $DB \le k-s$. Since $DB(h_1 + dh_4) \le k-s$, we have $h_1 + dh_4 = 0$. Therefore

$$h = dh_3 + (a_1 \otimes 1)h_4 - (1 \otimes x_1)dh_4,$$
$$= d(h_3 + (1 \otimes x_1)h_4),$$

so h is a coboundary.

(2)$\Longrightarrow$(3): trivial.

(3)$\longrightarrow$(1): first we shall show that (3) is also fulfilled in J_1'. Let h' be a cocycle in J_1' with exterior degree 1; we have to show that h' is a coboundary. We write

$$h' = \sum_{i=2}^{m} b_i' \otimes x_i, \quad d'h' = \sum_{i=2}^{m} b_i'a_i' \otimes 1 = 0.$$

Thus $\sum_{i=2}^{m} b_i'a_i' = 0$, and hence $\sum_{i=2}^{m} b_i a_i = -b_1a_1$ for some $b_1 \in B$. Therefore $h = \sum_{i=1}^{m} b_i \otimes x_i$ is a cocycle in J of exterior degree 1. Since (3) is assumed in J it follows that h = du, where $u = (1 \otimes x_1)h_1 + h_2$, with $h_1, h_2 \in J_1$. Thus

$$h = du = -(1 \otimes x_1)dh_1 + (a_1 \otimes 1)h_1 + dh_2,$$

and hence

$$\sum_{i=2}^{m} b_i \otimes x_i = (a_1 \otimes 1)h_1 + dh_2.$$

Projecting into J' we get $h' = d'h'_2$, and the assertion is proved.

By our inductive assumption and the fact just proved it follows that the annihilator of the projection of a_i in $B/(a_1,\ldots,a_{i-1})$ vanishes up to degree k-s for $i = 2,3,\ldots,m$. To complete the proof of (1) it remains only to show that the annihilator of a_1 in B is zero up to degree k-s. We shall do this by induction on the degree t of B^t.

For t = 0 the assertion is trivial: Assume it for t-1 and let $b \in B^t$ be such that $ba_1 = 0$. Then $h = b \otimes x_1$ is a cocycle. By (3) we can write

$$h = du, \quad u = (1 \otimes x_1)h_1 + h_2,$$

where $h_1, h_2 \in B \otimes \Delta^{\dagger}(x_2,\ldots,x_m)$, $DBh_1 = DBh_2 = t-s$, and exterior degrees 1 and 2 respectively. Then

$$du = (a_1 \otimes 1)h_1 - (1 \otimes x_1)dh_1 + dh_2,$$

from which $b \otimes x_1 = -(1 \otimes x_1)dh_1$ and $(a_1 \otimes 1)h_1 + dh_2 = 0$. Then $d'h'_2 = 0$ so that $h'_2 = d'h'_3$ for some h'_3. Therefore we can write

$$h_2 = dh_3 + (a_1 \otimes 1)h_4$$

for some h_4, and hence

$$(a_1 \otimes 1)h_1 + (a_1 \otimes 1)dh_4 = (a_1 \otimes 1)(h_1 + dh_4) = 0.$$

By the inductive assumption on t-1 we have $h_1 = -dh_4$: hence

$$b = \pm dh_1 = \pm ddh_4 = 0$$

and (1) is proved. This completes the lemma.

Remark: By induction on the degree it is clear that if $B = (a_1,\ldots,a_m) + B^o$ then B is generated (as an algebra) by 1 and the a_i. Hence in order to show that $B = K[a_1,\ldots,a_m]$ it suffices to prove that the a_i are ann-free and that B $B = (a_1,\ldots,a_m) + B^o$.

To illustrate we shall now prove theorem 19.1 in the special case where G is an m-dimensional torus T. We have $H^*(T) = \Lambda(x_1,\ldots,x_m)$ where $d^o x_i = 1$ and the x_i are universally transgressive. As J we take

$$E_2 \approx H^*(B_T) \otimes \Lambda(x_1,\ldots,x_m);$$

then d_2 is given by $d_2(H^*(B_T) \otimes 1) = 0$ and $d_2(1 \otimes x_i) = (y_i \otimes 1)$. We first prove by induction that the y_i are ann-free. Suppose they are ann-free up to k but not to k+1. Then by (3) in lemma (19.2) there must be a d_2-cocycle $h \in E_2$ which is not a d_2-coboundary and is of the form

$$h = \sum b_i \otimes x_i, \quad d^o b_i = k+1-2 = k-1.$$

Since $DFh = d^o x_i = 1$, h is a permanent cocycle. Since E_∞ is trivial there must be an $s > 2$ and an element $u \in E_2$ such that $d_r k_r^2 u = 0$ for $r < s$ and $d_s k_s^2 u = k_s^2 h \neq 0$. But

$$DBu = DBh-s = k-1-s \leq k-4,$$

$$DFu = 1+s-1 = s>2,$$

which contradicts (2) of the lemma. Thus the y_i are ann-free. We therefore have $H_e^{p,q} = 0$ for $e > 0$ and so

$$E_3 \approx H^*(B_T,K)/(y_1,\ldots,y_m) \otimes 1,$$

$$E_\infty = E_3.$$

Since E_∞ is trivial we have $H^*(B_T,K) = H^o(B_T,K) + (y_1,\ldots,y_m)$. In view of the above remark we then have the desired result,

$$H^*(B_T,K) = K[y_1,\ldots,y_m].$$

The general result of 19.1 will be obtained as a consequence of theorem 19.4 (below). First however we shall prove a preliminary lemma. Consider a canonical spectral sequence (E_r), $r \geq 2$, over a field K with E_r anti-commutative (with respect to total degree.) We assume further

$$E_2 = B \otimes \Delta(x_1,x_2,\ldots), \quad E_2^{p,q} = B^p \otimes (\Delta(x_1,x_2,\ldots))^q,$$

(of finite type) and the x_i transgressive. Let $y_i \in B$ be an image of x_i by transgression; explicitly, if $s = d^o x_i$ then $k_{s+1}^2 = d_{s+1} k_{s+1}^2 y_i$. We introduce the following notation:

P : the vector space spanned by the x_i,

P^r : the vector space spanned by the x_i of degree r,

Q : the vector space spanned by the y_i,

Q^r : the vector space spanned by the y_i of degree r,

$P_*^r = \sum_{j \geq r} P^j$, $Q_*^r = \sum_{j \geq r} Q^j$.

$B_r = k_r^2(B \otimes 1) = \sum_p E_r^{p,0}$.

Note that $B_r \otimes \Delta^\dagger P_*^{r-1}$ has a natural differential d'_r defined by

$d'_r(B_r \otimes 1) = 0$,

$d'_r(1 \otimes x_i) = 0$ if $x_i \in P_*^r$,

$d'_r(1 \otimes x_i) = k_r^2 y_i \otimes 1$ if $x_i \in P^{r-1}$.

The elements of $B \otimes \Delta^\dagger P_*^{r-1}$ are d_i-cocycles for $i < r$; hence $T_r = k_r^2(B \otimes \Delta^\dagger P_*^{r-1})$ is well defined and is a d_r-invariant subspace of E_r.

Lemma 19.3. Assume $B^1 = E_2^{1,0} = 0$, d_r is an isomorphism of P^{r-1} onto $k_r^2 Q^r$ for $r \leq i$, and that $Q^2, \ldots, Q^i$ are ann-free up to k. Then for $2 \leq r \leq i$ we have

(1_r) $H(B_r \otimes \Delta^\dagger P_*^{r-1}) \subset E_{r+1}$ for $DB \leq k+1$,

(2_r) $H(B_r \otimes \Delta^\dagger P_*^{r-1}) = B_r/(Q^r) \otimes \Delta^\dagger P_*^r$,

$\qquad \approx B_{r+1} \otimes \Delta^\dagger P_*^r = E_{r+1}$ for $DB \leq k-r$,

(3_r) $B_{r+1} \otimes \Delta^\dagger P_*^{r-1} = T_{r+1}$ for $DB \leq k+1$.

Proof. We proceed by induction on r.

(1_2) Clear.

(2_2) We can write

$$E_2 = B \otimes \Delta^\dagger P_*^1 = B \otimes \Delta^\dagger P^1 \otimes \Delta^\dagger P_*^2,$$

and by the Künneth rule

$$E_3 = H(B \otimes \Delta^\dagger P^1) \otimes \Delta^\dagger P_*^2.$$

The desired result now follows readily from lemma 19.2.

(3_2) We have

$$T_3 = k_3^2(B \otimes \Delta^\dagger P_*^2) = (B \otimes \Delta^\dagger P_*^2)/(B \otimes \Delta^\dagger P_*^2) \cap d_2E_2,$$

whence (3_2).

Now assume $(1_{r-1}),(2_{r-1}),(3_{r-1})$ for $r \leq i$. The map $k_r^2 : B \otimes \Delta^\dagger P_*^{r-1} \to T_r$ induces a map $f_r : B_r \otimes \Delta^\dagger P_*^{r-1} \to T_r$ which is surjective, and $f_r d'_r = d_r f_r$. We assert

(α) $T_r \cap d_rE_r = T_r \cap d_rT_r$ for $DB \leq k$,

that is to say, if a cocycle of T_r of $DB \leq k+1$ bounds in E_r it bounds in T_r. For if $h \in T_r$, $DBh \leq k+1$, $h = d_ru$, then $DBu \leq k-r+1$. But by (2_{r-1}) and (3_{r-1}), $E_r = T_r$ for $DB \leq k-r+1$; hence $u \in T_r$.

(1_r) $f_r^*\colon H(B_r \otimes \Delta^{\dagger}P^{r-1}) \to E_{r+1}$.

If h is a cocycle in $B_r \otimes \Delta^{\dagger}P_*^{r-1}$ with $DBh \leq k+1$ then $f_r(h)$ is a cocycle in T_r and hence in E_r. Suppose $f_r(h) = d_ru$; then $u \in T_r$ and $DBu \leq k-r+1$. By (3_{r-1}) h is then a coboundary in $B_r \otimes \Delta^{\dagger}P_*^{r-1}$. Therefore f_r^* is injective up to $DB \leq k+1$ and (1_r) is proved.

(2_r) The first equality follows from lemma 19.2, the second from 19.2 using also (α) and (2_{r-1}), and the third from (2_{r-1}) using the fact that E_{r+1} is a quotient of a subspace of E_r.

(3_r) We have

$$T_{r+1} = k_{r+1}^r k_r^2(B \otimes \Delta^{\dagger}P_*^r) = k_{r+1}^r T_r = k_{r+1}^r(B_r \otimes \Delta^{\dagger}P_*^r).$$

Using (α) we have the desired result.

Theorem 19.4. Let (E_r), $r \geq 2$, be a canonical spectral sequence over a field K with E_r anti-commutative. Suppose

$$E_2 = B \otimes \Delta^{\dagger}(x_1,\dots,x_i,\dots)$$

where the x_i are transgressive, have odd degree if the characteristic of K is not 2, and only finitely many of the x_i have the same degree. Assume further that E_∞ is trivial, $E_\infty^{0,0} = K, E^{p,q} = 0$ if $p+q > 0$). Let y_i be an image of x_i by transgression. Then

$$B = K[y_1,\dots,y_i,\dots].$$

As a corollary we have at once theorem 19.1 stated at the beginning of the section. Serre has also made use of the theorem in certain computations on the Eilenberg-MacLane groups.

Proof of theorem 19.4.

(α) We have $B^1 = E_2^{1,0} = E_\infty^{1,0} = 0$. Moreover for $r \geq 2$ the kernel of d_r on $E_r^{0,r-1}$ is $E_{r+1}^{0,r-1} = E_\infty^{0,r-1} = 0$. Thus d_r is injective on $E_r^{0,r-1} \supset P^{r-1}$ and so τ maps P^{r-1} isomorphically onto Q^r.

(β) We assert that the y_i are ann-free. Assume that the y_i are ann-free up to k and that $o = Q^1, Q^2, \dots, Q^i$ are ann-free up to k+1 for some $i \geq 2$. We shall prove that $Q^1, \dots, Q^{i+1}$ are ann-free up to k+1. Suppose this is not true. By (1_i) we have

$$H(B_i \otimes \Delta^{\dagger}P_*^{i-1}) \subset E_{i+1} \quad \text{for } DB \leq k+1$$

Using the Künneth rule and the definition of B_i we can write

$$H(B_i \otimes \Delta^{\dagger}P_*^{i-1}) = H(B/Q^1+\dots+Q^{i-1}) \otimes \Delta^{\dagger}P_*^{i-1}) \otimes \Delta^{\dagger}P_*^{i}$$

to which we can then apply lemma 19.2. Having assumed $Q^1,\dots,Q^{i+1}$ not ann-free up to k+1 it follows from (3) in the lemma that there is a d_{i+1}-cocycle $h \in E_{i+1}$ which is not a d_{i+1}-coboundary and is such that $DBh = k+1-i$, $DFh = i-1$. Because of its low fiber degree h is a permanent cocycle. Since E_∞ is trivial there is an $s \geq i + 1$ and an element $u \in E_2$ such that $d_r k_r^2 u = 0$ for $r < s$ and $d_s k_s^2 u = k_s^{i+1} h \neq 0$. We have

$$DBu = DBh-s = k+1-i-s < k-s+1$$

hence by (2_{s-1}) in lemma 19.3, we have

$$k_s^2 u \in B_s \otimes \Delta^{\dagger}P_*^{s-1}.$$

But also

$$DFu = DFh+s-1 = i-1+s-1 = i+s-2<2(s-1);$$

hence $k_s^2 u$ is not decomposable and

$$k_s^2 u \in B_s \otimes P_*^{s-1}.$$

Since d_s vanishes on transgressive elements with degree greater than s-1 and $d_s k_s^2 u \neq 0$ we must have

$$k_s^2 u \in B_s \otimes P^{s-1}.$$

Thus $s-1 = i+s-2$, implying $i = 1$ in contradiction of the fact that $i \geq 2$. This completes the induction, and the $Q^1,Q^2,\dots$ are ann-free.

Having established (α) and (β) we can now apply lemma 19.3 for all k. Thus for any k we have

$$B_{k+1}^k = B^k/(Q^1+\dots+Q^k) \cap B^k$$

from which we see that $B = B^0 + (Q^1 + Q^2+\dots)$. Since the y_i are ann-free we have the desired result $B = K[y_1,\dots,y_m,\dots]$ by the remark following 19.2.

20. Invariants of the Weyl group, and classifying spaces. The Hirsch formula.

Let (E,B,G) be a principal fibering with G a compact Lie group. Let U be a closed connected subgroup of G: then we also have a principal fibering (E,E/U,U). If now N is a subgroup of G which is contained in the normalizer of U then N operates on U by the inner automorphisms $u \to nun^{-1}$ and, since $nU = Un$, $(n \in N)$, it permutes the cosets xU of E and operates on the fibering (E,E/U,U). Therefore N operates on the cohomology of E, the cohomology of E/U, and the spectral sequence of (E,E/U,U). We have the following properties:

(a) $N/N \cap U$ acts on $H^*(E,A)$ and $H^*(E/U,A)$, the operations being compatible with the filtration and the map induced by the fiber map.

(b) the operations of N on

$$E_2 = H^*(E/U,\ H^*(U,A))$$

commute with the canonical isomorphisms

$$E_2^{p,o} = H^p(E/U,A), \qquad E_2^{o,q} = H^q(U,A).$$

In particular $N \cap U$ operates trivially so that $N/N \cap U$ operates on E_r, $r \geq 2$.

(c) the operations of $N/N \cap U$ on E_∞ agree with those induced by the operations on $H^*(E,A)$. In particular if A is a field K then there is a vector space isomorphism between $H^*(E,K)$ and E_∞ which commutes with the operations.

We can also consider the fibering $(E/U,B,G/U)$, (G/U the left coset space). Then N operates on G/U by right translations, on E/U, the operations being compatible with the fibering, and trivially on B. Therefore N operates on the cohomology of G/U, and, since U is connected, $N/N \cap U$ operates on $H^*(G/U,A)$. In addition we have the following properties:

(a') $N/N \cap U$ operates on $H^*(E/U,A)$ and trivially on $H^*(B,A)$, the operations being compatible with the filtration and the map induced by fiber map.

(b') Assuming G connected then N operates on the spectral sequence (E_r) of $(E/U,E/G,G/U)$, the operations on

$$E_2 = H^*(B,H^*(G/U,A))$$

commuting with the canonical isomorphisms

$$E_2^{p,o} = H^p(B,A), \quad E_2^{o,q} \approx H^q(G/U,A).$$

Since $N \cap U$ operates trivially, $N/N \cap U$ operates on (E_r), $r \geq 2$.

(c') The operations on E_∞ agree with those induced by the operations of $N/N \cap U$ on $H^*(E/U,A)$. If $A = K$ then there is a vector space isomorphism between $H^*(E/U,K)$ and E_∞ which commutes with the operations.

Let G be a compact Lie group then it contains maximal toral subgroups. Any two of these are conjugate under an inner automorphism of G, their common dimension is called the <u>rank</u> of G. Let T be a maximal torus and let N_T be the normalizer of T in G. It is known that T has finite index in N_T so that $W(G) = N_T/T$ is a finite group which is called the <u>Weyl group</u> of G. The map $t \to ntn^{-1}$, $n \in N_T$, $t \in T$, is an automorphism of T which depends only on the coset mod T in which t lies and induces a map $W(G) \to Aut(T)$ which is faithful when G is connected.

In the preceding discussion we take $U = T$ and $N = N_T$ and consider the principal fibering ($E_G = E_T, B_T, T$). Then W(G) operates in accordance with (a),(b),(c) above. Now $T = S_1 \times \cdots \times S_1$, (m times), has no torsion so that

$$E_2 \approx H^*(B_T,Z) \otimes H^*(T,Z).$$

Using the acyclicity of E_G and the fact that $d_r \equiv 0$, $r \geq 3$, it follows that the transgression τ maps $H^1(T,Z)$ isomorphically onto $H^2(B_T,Z)$. By theorem 19.1

$$H^*(B_T,Z) = Z[v_1,\ldots,v_m], \quad v_i \in H^2(B_T,Z),$$

where v_i is an image of a generator of $H^1(T,Z)$ by transgression. The identification of $H^*(B_T,Z)$ with polynomials over $H^1(T,Z)$ by τ is easily seen to commute with the operations of W(G).

The ring of invariants I_G of W(G) in $H^*(B_T,Z)$ is a direct summand. For if $kx \in I_G$ and $w \in W(G)$ then $kx = w(kx = kw(x)$, and since $H^*(B_T,Z)$ has no torsion it follows that $x = w(x)$. Then

$$I_G \otimes Z_p \subset H^*(B_T,Z) \otimes Z_p = H^*(B_T,Z_p).$$

We shall now study the second fibering ($B_T,B_G,G/T$).

<u>Proposition 2o.1</u>. If G is a compact connected Lie group and T a maximal torus then G/T has no torsion, its odd dimensional Betti numbers are zero, and $\chi(G/T)$ is equal to the order of W(G).

For the proof of 2o.1 in the case of the classical groups and G_2,F_4, see [1] and in the general case [3],[4].

<u>Proposition 2o.2</u>. (Leray). The natural representation Γ_w of W(G) in $H^*(G/T,R)$ is equivalent to the regular representation.

<u>Proof.</u> Let T_n be the homeomorphism of G/T defined by $g \to g\cdot n$. Clearly if T_n has a fixed point then $n \in T$. Let $w \in W(G)$, then for the Lefschetz number L(w) we have

$$\begin{aligned} L(w) &= \sum (-1)^i \operatorname{tr}\Gamma_w \text{ in } H^i(G/T,R), \\ &= \sum \operatorname{tr}\Gamma_w \text{ in } H^i(G/T,R), \\ &= \quad \operatorname{tr}\Gamma_w \text{ in } H^*(G/T,R). \end{aligned}$$

If $w \neq e$ then $L(w) = 0$; hence $\operatorname{tr}\Gamma_w = 0$. On the other hand if $w = e$ then $\operatorname{Tr}\Gamma_w = \dim H^*(G/T,R) =$ order of W(G). This is precisely the character of the regular representation.

As a corollary we see that the trivial representation occurs exactly once (in $H^0(G/T,R)$). Since W(G) operates on $H^*(G/T,Z_0)$ and since

$$H^*(G/T,K_0) = H^*(G/T,Z_0) \otimes K_0,$$

this applies also to $H^*(G/T,K_0)$.

Theorem 2o.3. Let G be a compact connected Lie group with no p-torsion and let T be a maximal torus. Then

(a) $\rho_p^*(T,G) : H^*(B_G,K_p) \to H^*(B_T,K_p)$ is an isomorphism onto $I_G^+ \otimes K_p$;

(b) $H^*(G/T,K_p)$ is the characteristic ring of the fibering of G by T and

$$H^*(G/T,K_p) = H^*(B_T,K_p)/(I_G^+ \otimes K_p),$$

(I_G^+ is generated by elements of positive degree).

Remark. A similar statement holds over Z when G has no torsion.

Proof. We consider the fibering $(B_T,B_G,G/T)$; we have

$$E_2 \approx H^*(B_G,K_p) \otimes H^*(G/T,K_p),$$

$$E_\infty \approx \mathrm{Gr}H^*(B_T,K_p).$$

The second factor in E_2 has only even dimensions as a consequence of proposition 2o.1. Since G has no p-torsion we have

$$H^*(G,K_p) = \Lambda(x_1,\ldots,x_m),\ d^o x_i \text{ odd}.$$

Therefore by the theorem on transgression (19.1), $H^*(B_G,K_p)$ has only even dimensions. Thus E_2 has only even dimensions and it follows that $E_2 = E_\infty$. This proves (b) and that $\rho_p^*(T,G)$ is injective. It remains to determine the image $\rho_p^*(T,G)\ H^*(B_G,K_p) \subset H^*(B_T,K_p)$.

Case p = o. We know W(G) operates trivially on $H^*(B_G,K_p)$,(see (a') at beginning of the section). By (2o.2) it acts as the regular representation on $H^*(G/T,K_o)$ with $H^o(G/T,K_o)$ as the only trivial representation. This implies that

$$H^*(B_G,K_o) \otimes H^o(G/T,K_o)$$

is the set of all invariants of W(G) in E_2. Its projection in E_∞ is then the image of $\rho_p^*(T,G)$. We have already proved that $E_2 = E_\infty$. Using the vector space isomorphism in (c') we can identify $E_\infty = H^*(B_T,K_o)$, and hence also

$$H^*(B_G,K_o) \otimes H^o(G/T,K_o) = I_G \otimes K_o.$$

Thus the image of $\rho_p^*(T,G)$ is the full set of invariants $I_G \otimes K_o$.

Case p ≠ 0. Since I_G is a direct summand, dim $(I_G \otimes K_p)^i$ is equal to the rank of $(I_G)^i$, and hence by the preceding to dim $H^i(B_G,K_o)$. Since G has no p-torsion, B_G has no p-torsion (see 18.4, 18.5); thus

$$\dim H^i(B_G,K_p) = \dim H^i(B_G,K_o).$$

Moreover

$$H^*(B_G,K_p) \approx H^*(B_G,Z) \otimes K_p,$$

whence

$$\rho_p^*(T,G) \approx \rho_Z^*(T,G) \otimes K_p$$

and the image Q of $\rho_p^*(T,G)$ is contained in $I_G \otimes K_p$. By (a') and the preceding remarks, $Q \cap H^i(B_T,K_p)$ and $(I_G \otimes K_p)^i$ have the same dimension and so must be equal.

Proposition 2o.4. If G is a compact Lie group and T a maximal torus then $\rho_o^*(T,G)$ maps $H^*(B_G,K_o)$ isomorphically onto $I_G \otimes K_o$.

Proof. Let G_o be the connected component of e, then T is maximal torus in G_o. We have natural projections

$$E_G/T \to E_G/G_o \to E_G/G,$$

that is,

$$B_T \xrightarrow{\rho(T,G_o)} B_{G_o} \xrightarrow{\lambda} B_G,$$

with $\rho(T,G) = \lambda \cdot \rho(T,G_o)$. The map λ is clearly a covering map with covering group G/G_o. Clearly $W(G_o)$ is an invariant subgroup of $W(G)$ and one sees readily that W(G)/W(G $W(G)/W(G_o) = G/G_o$. We have $\rho^*(T,G) = \rho^*(T,G_o) \cdot \lambda^*$. It is known that λ^* is injective with the invariants of the covering group in $H^*(B_{G_o}, K_o)$ as image. If we now apply theorem 2o.3 to $\rho^*(T,G_o)$ the proposition is proved.

Remark. If p is prime to the order of G/G_o it is easily seen that $I_G \otimes K_p$ is the set of all invariants of G/G_o in $I_{G_o} \otimes K_p$. Using this one proves as before that $\rho_p^*(T,G)$ maps $H^*(B_G,K_p)$ isomorphically onto $I_G \otimes K_p$ when G_o has no p-torsion and p is prime to the order of G/G_o.

Theorem 2o.5. Let G be a compact connected Lie group and let T be a maximal torus. If $2r_1-1,\ldots,2r_m-1$ are the degrees of primitive elements in $H^*(G,K_o)$ then m is also the rank of G, and

$$P(G/T,t) = \frac{(1-t^{2r_1})\ldots(1-t^{2r_m})}{(1-t^2)^m} .$$

Moreover, the order of $W(G) = \Pi r_i$ and the dimension of $G = 2 \sum r_i - m$.

Proof. We have seen earlier that

$$H^*(B_T,K_o) = K_o[v_1,\ldots,v_m], \quad d^o v_i = 2;$$

hence

$$P(B_T,t) = \frac{1}{(1-t^2)^m} .$$

By Hopf's theorem

$$H(G,K_o) = \Lambda(x_1,\ldots,x_m), \quad d^o x_i \text{ odd.}$$

Applying 18.1 we have

$$H^*(B_G,K_o) = K_o[y_1,\ldots,y_{m'}], \quad d^o y_i = 2r_i;$$

hence

$$P(B_G,t) = \frac{1}{(1-t^{2r_1})\cdots(1-t^{2r_{m'}})} .$$

We have already noted that in the spectral sequence of $(B_T, B_G, G/T)$ we have $E_2 = E_\infty$ since all degrees are even. Therefore

$$P(G/T,t) = P(B_T,t)/P(B_G,t),$$

$$= \frac{(1-t^{2r_1})\cdots(1-t^{2r_{m'}})}{(1-t^2)^m}$$

We assert that $m = m'$. If $m' < m$ then on dividing we have that $P(G/T,t)$ is an infinite series. But this contradicts the fact that $H^*(G/T,K_o)$ has finite dimension; hence $m' \geq m$. On the other hand if $m' > m$ then putting $t = 1$ in

$$P(G/T,t) = \prod_{i=1}^{m} (1+t^2+\ldots+ t^{2r_i-2}) \prod_{i=m+1}^{m'} (1-t^{2r_i})$$

gives $P(G/T,1) = 0$. This contradicts the fact that $P(G/T,1) > 0$; hence $m' = m$.

By proposition (2o.1) we have that the order of $W(G)$ is equal to $\chi_p(G/T) = P(G/T,-1) = P(G/T,1)$, the last equality holding since only even dimensional Betti numbers are present. Therefore

$$\text{order of } W(G) = \prod_{i=1}^{m} (1+t^2+\ldots+t^{2r_i-2}) \Big|_{t=1} = \prod_{i=1}^{m} r_i .$$

Since G/T is an orientable manifold we have

$$\dim G/T = \sum_{i=1}^{m} (2r_i-2) = 2\sum r_i-2m.$$

Thus $\dim G-m = 2\sum r_i-2m$, and we have the desired result.

Theorem 2o.6. Let G be a compact connected Lie group of rank m. Let U be a closed connected subgroup of the same rank. Let $2r_1-1,\ldots,2r_m-1$ be the degrees of the primitive generators of $H^*(G,K_o)$ and $2s_1-1,\ldots,2s_m-1$ the degrees of the primitive generators of $H^*(U,K_o)$.

(a) If U has no p-torsion then G/U has no p-torsion and we have the formula (conjectured by Hirsch for $p = 0$):

$$P_p(G/U,t) = \frac{(1-t^{2r_1})\cdots(1-t^{2r_m})}{(1-t^{2s_1})\cdots(1-t^{2s_m})} .$$

(b) If G and U have no p-torsion then $H^*(G/U,K_p)$ is equal to its characteristic subring, and

$$H^*(G/U,K_p) \approx (I_U \otimes K_p) \,/\, (I_G^+ \otimes K_p).$$

Proof. (a) Consider the spectral sequence of $(G/T,G/U,U/T,\pi)$ defined by $G \supset U \supset T$. We have

$$E_2 \approx H^*(G/U,K_p) \otimes H^*(U/T,K_p),$$
$$E_\infty \approx GrH^*(G/T,K_p).$$

Since U has no p-torsion then by (2o.3) $H^*(U/T,K_p)$ is equal to the characteristic ring. By corollary (16.4) U/T is totally non-homologous to 0 in $(G/T,G/U,U/T,\pi)$ (mod p). Therefore $E_2 = E_\infty$ and

$$P_p(G/U,t)\cdot P_p(U/T,t) = P_p(G/T,t).$$

Applying (2o.5) the Hirsch formula follows. Noting that in characteristics o and p we get the same P(G/U,t), we see that G/U has no p-torsion.

(b) Consider the spectral sequence of the fibering $(B_U,B_G,G/U,p(U,G))$: we have

$$E_2 = H^*(B_G,K_p) \otimes H^*(G/U,K_p),$$
$$E_\infty = GrH^*(B_U,K_p).$$

By(19.1) and the Hirsch formula the factors in E_2 have only even degrees and it follows that $E_2 = E_\infty$. Therefore G/U is totally non-homologous to 0 which means that i^* is surjective. But i^* is the characteristic map as is shown in section 17. This proves that $H^*(G/U,K_p)$ is the characteristic ring. If we also apply theorem 14.2 we get

$$H^*(G/U,K_p) \approx H^*(B_U,K_p) \;/\; (\rho_p^*(U,G)H^+(B_G,K_p)).$$

Applying $\rho_p^*(T,U)$, which by 2o.3(a) is injective, we get

$$\begin{aligned} H^*(G/U,K_p) &\approx I_U \otimes K_p/ \; (\rho_p^*(T,U)\rho_p^*(U,G)\; H^+(B_G,K_p)),\\ &= I_U \otimes K_p/ \; (\rho_p^*(T,G)\; H^+(B_G,K_p)),\\ &= I_U \otimes K_p/ \; (I_G^+ \otimes K_p), \end{aligned}$$

the last equality holding by 2o.3(a).

Remark: We have a similar statement over Z: if G and U have no torsion and G/U has no torsion then

$$H^*(G/U,Z) \approx I_U/(I_G^+).$$

BIBLIOGRAPHY

[1] A. Borel, Ann. of Math. 57(1953), 115-2o7.

[2] A. Borel, Amer. J. Math. 76 (1954), 273-342.

[3] A. Borel, Kählerian coset spaces of semisimple Lie groups, Proc.Nat.Acad.Sci., Vol. 4o, No. 12, (1954), 1147-1151.

[4] R. Bott, On torsion in Lie groups,Proc.Nat.Acad.Sci.,Vol.4o,No.7 (1954),586-588.

[5] H. Cartan, Séminaire, Paris (1949-5o).

[6] N. Steenrod, The topology of fibre bundles, Princeton (1951).

CHAPTER IV

CLASSIFYING SPACES OF THE CLASSICAL GROUPS

21. Unitary groups.

We shall use the following notation:

$S(a_1,\dots,a_r)$: the ring of symmetric functions in $a_1,\dots,a_r$,

$S^+(a_1,\dots,a_r)$: the elements of positive degree in $S(a_1,\dots,a_r)$,

σ_i : the ith elementary symmetric function.

Let $U(n)$ denote the group of $n \times n$ unitary matrices and $W_{n,s}$ the complex Stiefel manifold of orthonormal s-frames in n-dimensional Hermitian space. Evidently $W_{n,1} = S_{2n-1}$ and $W_{n,n} = U(n)$. We identify $W_{n,s} = U(n)/U(n-s)$ in the usual way and let $\pi_{n,s} : U(n) \to W_{n,s}$ denote the natural map. More generally we may consider

$U(n) \supset U(n-t) \supset U(n-s)$, $t \le s$;

then the natural map $\pi_{s,t} : W_{n,s} \to W_{n,t}$ is a fibre map for the fibering $(W_{n,s},W_{n,t},W_{n-t,s-t})$. Moreover if $u \le t \le s$ then clearly $\pi_{s,u} = \pi_{t,u}\cdot\pi_{s,t}$.

Proposition 21.1. $H^*(W_{n,s},Z) = \Lambda(x^{(s)}_{2n-1},x^{(s)}_{2n-3},\dots,x^{(s)}_{2(n-s)+1})$ where the subscripts denote the degrees of the elements. Moreover the elements

$$\pi^*_{s,t}(x^{(t)}_{2j+1}) = x^{(s)}_{2j+1}\ , \quad (n-t \le j \le n-1)$$

generate the image of $\pi^*_{s,t}$.

Proof. For $s = 1$ the proposition is clear. Assume it for $s-1$, $(s>1)$, and consider the fibering $(W_{n,s},W_{n,s-1},W_{n-s+1,1},\pi_{s,s-1})$. We have

$$E_2 \approx H^*(W_{n,s-1},Z) \otimes H^*(W_{n-s+1,1},Z);$$

then, using the inductive assumption and $W_{n-s+1,1} = S_{2(n-s+1)-1}$,

$$E_2 = \Lambda(x^{(s-1)}_{2n-1},\dots,x^{(s-1)}_{2(n-s+1)+1}) \otimes \Lambda(x^{(1)}_{2(n-s)+1}).$$

One shows readily by an argument on the fibre degrees of elements in E_r that $d_r \equiv 0$ for $r \ge 2$; hence

$$E_2 = E_\infty = GrH^*(W_{n,s},Z).$$

By the analogous statements to 14.1. and 14.2 (a) for integer coefficients we have that i^* is surjective and $\pi^*_{s,s-1}$ injective. Consider the elements

$$\pi^*_{s,s-1}(x^{(s-1)}_{2(n-i)+1}) = x^{(s)}_{2(n-i)+1},\quad i = 0,1,\dots,s-1,$$

and an element $x^{(s)}_{2(n-s)+1}$ such that

$$i^*\ x^{(s)}_{2(n-s)+1} = x^{(1)}_{2(n-s)+1}.$$

These are clearly linearly independent and generate $H^*(W_{n,s},Z)$. Since they also have odd degrees and $H^*(W_{n,s},Z)$ is torsion free it follows that

$$H^*(W_{n,s},Z) = \Lambda(x^{(s)}_{2n-1},\dots,x^{(s)}_{2(n-s)+1}).$$

This also proves the second statement in the theorem in the case where $t = s-1$. The general statement then follows easily by induction.

Proposition 21.2. The elements $x^{(s)}_{2n-1},\dots,x^{(s)}_{2(n-s)+1}$ are universally transgressive.[f]

Proof. If $s=1$, $W_{n,s}$ is a sphere so the generator $x^{(1)}_{2n-1}$ is clearly universally transgressive. Assume the proposition for $s-1$, $(s > 1)$; then $x^{(s-1)}_{2n-1},\dots,x^{(s-1)}_{2(n-s+1)+1}$ are transgressive in the fibering

$$(E_{U(n)}/U(n-s+1),B_{U(n)},W_{n,s-1}).$$

The natural map

$$\lambda : E_{U(n)}/U(n-s) \to E_{U(n)}/U(n-s+1)$$

defines a representation of $(E_{U(n)}/U(n-s),B_{U(n)},W_{n,s})$ in the preceding fibering such that $\bar{\lambda} : B_{U(n)} \to B_{U(n)}$ is the identity map. If we canonically identify the standard fibres $W_{n,s}$ and $W_{n,s-1}$ with the various fibres in $E_{U(n)}/U(n-s)$ and $E_{U(n)}/U(n-s+1)$ respectively, then λ restricted to a fibre is precisely the map $\pi_{s,s-1} : W_{n,s} \to W_{n,s-1}$. It follows from this that the elements

$$x^{(s)}_{2n-i} = \pi^*_{s,s-1}(x^{(s-1)}_{2n-i}),\quad i = 1,3,\dots,2s-3,$$

are transgressive in $(E_{U(n)}/U(n-s),B_{U(n)},W_{n,s})$. The element $x^{(s)}_{2(n-s)+1}$ having lowest possible degree is clearly also transgressive.

Theorem 21.3. $B_{U(n)}$ has no torsion and

$$H^*(B_{U(n)},Z) = Z[y_2,y_4,\dots,y_{2n}]$$

where the y's are images of the $x^{(n)}_{2n-i}$ by transgression.

This follows from 18.5(a) and the analogue of 18.1 for integers.

The unitary group $U(n)$ clearly has rank n; a maximal torus T^n consists of the diagonal matrices of the form

$$\begin{pmatrix} e^{2\pi i x_1} & & 0 \\ & \ddots & \\ 0 & & e^{2\pi i x_n} \end{pmatrix}$$

(f) In analogy with the definition of section 18 we say in the case of associated bundles that $x \in H^*(F)$ is <u>universally transgressive</u> if it is transgressive in the bundle $((E_G,F)_G,B_G,F)$ introduced in section 17.

The Weyl group $W(U(n))$ is then the group of permutations of the diagonal terms. We have shown in Chapter III that $H^*(B_{T^n},Z)$ may be identified by transgression with the ring of of polynomials over $H^1(T^n,Z)$, the operations $W(U(n))$ being compatible with the identification. Thus

$$H^*(B_{T^n},Z) = Z[v_1,\dots,v_n],\ d^o v_i = 2.$$

The Weyl group is then the permutation group of $(v_1,\dots,v_n)$ and $I_{U(n)} = S(v_1,\dots,v_n)$.

Theorem 21.4. $\rho^*(T^n, U(n))$ maps $H^*(B_{U(n)},Z)$ isomorphically onto $S(v_1,\dots,v_n)$.

This follows from the analogue of 20.3(a) for integers.

Given a fibering $(E,B,W_{n,n-i+1})$ we define the Chern class $C_{2i} \in H^{2i}(B,Z)$ as the image by transgression of the generator $x_{2i-1}^{(n-i+1)}$ in $H^{2i-1}(W_{n,n-i+1},Z)$. It is uniquely determined because by 21.2 the differentials d_r of the spectral sequence are identically zero for $2 \leq r < 2i$. (Usually C_{2i} is defined as the obstruction to extension of a cross-section in the given bundle, but this is equivalent to the above definition (see 37.16 in [7]). The image $C_{2i} \in H^{2i}(B_{U(n)},Z)$ of $x_{2i-1}^{(n-i+1)}$ by transgression in the fibering $(E_{U(n)}/U(i-1),B_{U(n)},W_{n,n-i+1}) = (B_{U(i-1)},B_{U(n)},W_{n,n-i+1})$ is the universal Chern class. Thus theorem 21.3 asserts $H^*(B_{U(n)},Z) = Z[C_2,C_4,\dots,C_{2n}]$. If $\phi : B \to B_{U(n)}$ is a classifying map for a fibering $(E,B,W_{n,n-i+1})$ then the image of the universal Chern class C_{2i} under ϕ^* is clearly the corresponding Chern class in the latter fibering.

Proposition 21.5. The universal Chern class C_{2i} generates the kernel of $\phi^*(U(i-1),U(n))$ in dimension $2i$.

Proof. In the spectral sequence of $(B_{U(i-1)},B_{U(n)},W_{n,n-i+1})$ we have

$$\begin{aligned} E_2 &= H^*(B_{U(n)},Z) \otimes H^*(W_{n,n-i+1},Z), \\ &= Z[y_2,y_4,\dots,y_{2n}] \otimes \Lambda(x_{2n-1}^{(n-i+1)},\dots,x_{2i-1}^{(n-i+1)}); \\ E_\infty &= GrH^*(B_{U(i-1)},Z) = GrZ[\bar{y}_2,\bar{y}_4,\dots,\bar{y}_{2i-2}], \\ &= Z[\bar{y}_2,\ \bar{y}_4,\dots,\bar{y}_{2i-2}] \end{aligned}$$

since there are only even degrees. Clearly then $\rho^*(U(i-1), U(n))$maps $Z[y_2,y_4,\dots,y_{2i-2}]$ isomorphically onto $H^*(B_{U(i-1)},Z)$. We can write

$$H^{2i}(B_{U(n)},Z) = D^{2i} + Z[y_{2i}]$$

where D^{2i} consists of the decomposable elements of degree $2i$. Thus the kernel of $\rho^*(U(i-1),U(n))$ in dimension $2i$ is $Z[y_{2i}] = Z[C_{2i}]$.

Proposition 21.6. $\rho^*(T^n,U(n))$ maps C_{2i} onto $\pm\sigma_i$.

We first make a general remark on $\rho^*(U,G)$ where U is a closed connected subgroup of G. Suppose we choose maximal tori in U and G respectively such that $T^s \subset T^n$. Correspondingly we have a commutative diagram

$$\begin{array}{ccc} H^*(B_{T^s},A) & \longleftarrow & H^*(B_{T^n},A) \\ \uparrow & & \uparrow \\ H^*(B_U,A) & \longleftarrow & H^*(B_G,A) \end{array}$$

where the indicated maps are induced by the corresponding ρ maps. Suppose now that the vertical maps are injective as will be the case in the proposition we wish to prove. Then instead of $\rho^*(U,G)$ we may consider $\rho^*(T^s,T^n)$ restricted to the image of $H^*(B_G,A)$ in $H^*(B_{T^n},A)$.

Proof of 21.6. If we regard $U(s) \subset U(n)$ in the usual way then we have a natural embedding $i : T^s \subset T^n$. The matrices of T^n are diagonal matrices whose diagonal elements are $e^{2\pi i x_1},\dots,e^{2\pi i x_n}$. Then the diagonal elements of the matrices of T^s are $1,\dots 1,\dots,1,e^{2\pi i \bar{x}_{n-s+1}},\dots,e^{2\pi i \bar{x}_n}$. One sees readily that we may regard the x_j and $\bar{x}_k$ as generators of $H^1(T^n,Z)$ and $H^1(T^s,Z)$ respectively; hence the map

$$i^* : H^1(T^n,Z) \to H^1(T^s,Z),$$

is given by

$$i^* x_j = \begin{cases} 0 & \text{if } j \leq n-s, \\ \bar{x}_j & \text{if } j > n-s. \end{cases}$$

We may write

$$H^*(B_{T^n},Z) = Z[v_1,\dots,v_n], \quad v_j = \tau(x_j),$$
$$H^*(B_{T^s},Z) = Z[\bar{v}_{n-s+1},\dots,\bar{v}_n], \quad \bar{v}_j = \tau(\bar{x}_j),$$

in which case the map $\rho^*(T^s,T^n)$ is given by

$$\rho^*(T^s,T^n)(v_j) = \begin{cases} 0 & \text{if } j \leq n-s, \\ v_j & \text{if } j > n-s. \end{cases}$$

We want to prove $\rho^*(T^n,U(n))$ maps $C_{2(s+1)}$ onto $\pm\sigma_{s+1}$ for any s. By the preceding proposition we know that $C_{2(s+1)}$ generates the kernel of $\rho^*(U(s),U(n))$ in dimension $2(s+1)$. On applying the preceding general remarks here we see that $\rho^*(T^n,U(n))(C_{2(s+1)})$ is a symmetric function of the v_j of degree $s+1$ which vanishes when $v_1,\dots,v_s$ are set equal to zero; hence (by a theorem on elementary symmetric functions) it follows that

$$\rho^*(T^n,U(n))(C_{2(s+1)}) = kv_1 \cdots v_{s+1}.$$

Since $C_2,\dots,C_{2(s+1)}$ generate $H^{2(s+1)}(B_{U(n)},Z)$ we must have $k = \pm 1$.

Remarks. We shall not discuss the various sign conventions for the Chern classes which are necessary for so called "duality" formulas. Briefly, these are such that

$$\rho^*(T^n,U(n))(C_{2i}) = \sigma_i$$

for suitable v_i permuted by $W(U(n))$. We also recall the "duality" formulas:

Given two principal bundles $(E^{(i)},B,U(n_i))$, $(i = 1,2)$, over B we construct a bundle $(E,B,U(n_1) \times U(n_2))$ over B by taking the bundle induced on the diagonal of $B \times B$ by the bundle

$$(E^{(1)} \times E^{(2)}, B \times B, U(n_1) \times U(n_2)).$$

Extending the structural group to $U(n)$, $n = n_1 + n_2$, we then have a bundle $(E^{(3)},B,U(n))$ which may be called the "Whitney sum" of the original two bundles. An associated bundle of the sum bundle with fibre C^n is obtained from associated bundles of the original bundles with fibres C^{n_1} and C^{n_2} respectively by taking as fibre the direct sum of the fibres.

Now let $C_{2j}^{(i)}$, $(i = 1,2,3)$ be the corresponding j^{th} - Chern Classes; then we have the "duality" formulas [4],

(21.7)
$$C_{2j}^{(3)} = \sum_{\alpha+\beta=j} C_{2\alpha}^{(1)}\cdot C_{2\beta}^{(2)} .$$

As is well known (and readily seen) this is in fact a relation among the universal Chern classes. For consider the standard inclusion $U(n_1) \times U(n_2) \subset U(n)$; this induces a map

$$\phi^*: H^*(B_{U(n)},Z) \to H^*(B_{U(n_1)},Z) \otimes H^*(B_{U(n_2)},Z)$$

(using the Künneth rule and the absence of torsion). Thus duality means:

(21.8)
$$\phi^* \quad (C_{2j}^{(3)}) = \sum_{\alpha+\beta=j} C_{2\alpha}^{(1)} \otimes C_{2\beta}^{(2)}.$$

Now a maximal torus T^n of $U(n)$ may be regarded as a product $T^{n_1} \times T^{n_2}$ where T^{n_i} is a maximal torus of $U(n_i)$. Using the preceding results we see that (21.8) is a translation of the elementary identity:

(21.9)
$$\sigma_j(y_1,\ldots,y_{n_1},z_1,\ldots,z_{n_2}) = \sum_{\alpha+\beta=j} \sigma_\alpha(y_1,\ldots,y_{n_1})\cdot\sigma_\beta(z_1,\ldots,z_{n_2}).$$

22. Orthogonal groups, cohomology mod 2.

Let $O(n)$ and $SO(n)$ denote the groups of $n \times n$ orthogonal and special orthogonal matrices respectively, and let $V_{n,s}$ denote the Stiefel manifold of orthonormal s-frames in n-dimensional Euclidean space. Clearly

$$V_{n,1} = S_{n-1} , V_{n,n-1} = V_{n,n} = SO(n).$$

We embed $O(n-s) \subset O(n)$ in the usual way and identify

$$V_{n,s} = O(n)/O(n-s) = SO(n)/SO(n-s).$$

If $O(n) \supset O(n-t) \supset O(n-s)$ then the natural map $\pi_{s,t} : V_{n,s} \to V_{n,t}$ is a fibre map for the fibering $(V_{n,s}, V_{n,t}, V_{n-t,s-t})$. If $u \le t \le s$ then $\pi_{s,u} = \pi_{t,u} \cdot \pi_{s,t}$. We assume the following well known results:

Lemma 22.1.

(a) For n even, $H^*(V_{n,2},Z) = \Lambda\,(x^{(2)}_{n-1}, x^{(2)}_{n-2})$.

(b) For n odd, $H^\circ(V_{n,2},Z) = H^{2n-3}(V_{n,2},Z) = Z$,

$H^{n-1}(V_{n,2},Z_2) = Z_2$,

$H^i(V_{n,2},Z_2) = 0$ otherwise.

Corollary. For n odd:

(a) $H^*(V_{n,2},Z_p) = H^*(S_{2n-3},Z_p)$ if $p \ne 2$.

(b) $H^*(V_{n,2},Z_2) = \Lambda\,(x^{(2)}_{n-1}, x^{(2)}_{n-2})$ if $n > 3$.

(c) $V_{3,2} = SO(3) = P_3$; hence

$H^*(V_{3,2},Z_2) = Z_2[x]/(x^4)$.

Proposition 22.2.

(a) $H^*(V_{n,s},Z_2) = \Delta(x^{(s)}_{n-1}, x^{(s)}_{n-2}, \dots, x^{(s)}_{n-k})$, $\quad d^\circ x^{(s)}_j = j$.

(b) $\pi^*_{s,t}(x^{(t)}_{n-i}) = x^{(s)}_{n-i}$ for $1 \le i \le t \le s \le n-1$.

Proof. For any n, if $s = 1$ then $V_{n,1} = S_{n-1}$; hence the proposition is true for $s = 1$. Assume it true for n-1 and any s and also for n and any s-1; then we prove it for n and s as follows. The inclusions $O(n) \supset O(n-1) \supset O(n-s)$ define a natural fibering $(V_{n,s}, S_{n-1}, V_{n-1,s-1})$. In its spectral sequence we have

$$E_2 \approx H^*(S_{n-1},Z_2) \otimes H^*(V_{n-1,s-1},Z_2),$$

$$= \Lambda\,(\bar{x}^{(1)}_{n-1}) \otimes \Delta(\bar{x}^{(s-1)}_{n-3}, \dots, \bar{x}^{(s-1)}_{n-s}), \quad d^\circ \bar{x}^{(s-1)}_{n-i} = n-i,$$

by inductive assumption. We assert that $H^{n-s}(V_{n,s},Z_2) \ne 0$. For $s = 2$ this follows from lemma 22.1. If $s > 2$ one sees readily that $x^{(s-1)}_{n-s}$ is a non-trivial permanent cocycle and hence the assertion follows.

Now consider the spectral sequence of the fibering $(V_{n,s}, V_{n,s-1}, S_{n-s}, \pi_{s,s-1})$ which is given by the inclusions $O(n) \supset O(n-s+1) \supset O(n-s)$. We have

$$E_2 \approx H^*(V_{n,s-1},Z_2) \otimes H^*(S_{n-k},Z_2),$$

$$= \Delta\,(x^{(s-1)}_{n-1}, x^{(s-1)}_{n-2}, \dots, x^{(s-1)}_{n-s+1}) \otimes \Lambda(x^{(1)}_{n-s}).$$

Clearly $d_r \equiv 0$ if $r \ne n-s+1$ and we assert that $d_{n-s+1} \equiv 0$. If $d_{n-s+1} \ne 0$ then it must not vanish on $x^{(1)}_{n-s}$ since

$$ {}^{n-s}E_{n-k+1} = {}^{n-s}E_2 = E_2^{o,n-s} = \Lambda(x_{n-s}^{(1)}) $$

has one generator. But then ${}^{n-s}E_{n-k+2} = 0$ which in turn implies that ${}^{n-s}E_\infty = 0$ which contradicts the assertion $H^{n-s}(V_{n,s},Z_2) \neq 0$ proved above. Therefore $d_{n-s+1} \equiv 0$, and hence $E_2 = E_\infty$. We define

$$ x_{n-i}^{(s)} = \pi_{s,s-1}^*(x_{n-i}^{(s-1)}) \quad \text{for } i \leq s-1, $$

and $x_{n-s}^{(s)}$ as any element such that

$$ i^*(x_{n-s}^{(s)}) = x_{n-s}^{(1)}. $$

It follows readily from elementary results on the tensor product of simple systems that

$$ H^*(V_{n,s},Z_2) = \Delta(x_{n-1}^{(s)},\ldots,x_{n-s}^{(s)}). $$

This also proves (b) in the case where $t = s-1$; the general case follows easily by induction.

Proposition 22.3. The $x_{n-i}^{(s)}$, $(i \leq s)$ are universally transgressive.

The proof is analogous to the proof of (21.2). For $i \geq 2$ we define the ith universal Stiefel-Whitney class mod 2 to be the element w_i which is the image by transgression of $x_{i-1}^{(n-i+1)}$ in the bundle

$$ (E_{O(n)}/O(i-1),B_{O(n)},O(n)/O(i-1)) = (B_{O(i-1)},B_{O(n)},V_{n,n-i+1}), $$

or equivalently in the bundle,

$$ (E_{O(n)}/SO(i-1),B_{SO(n)},SO(n)/SO(i-1)) = (B_{SO(i-1)},B_{SO(n)},V_{n,n-i+1}). $$

Thus w_i denotes an element of $H^i(B_{O(n)},Z_2)$ or of $H^i(B_{SO(n)},Z_2)$; however no confusion will arise since the map $\rho^*(SO(n),O(n))$ obviously maps the first onto the second. (Note that $\rho(SO(n),O(n))$ is actually a covering since it has a discrete fibre Z_2.) Finally we define the Stiefel-Whitney class mod 2, w_1 to be the non-zero element of $H^1(B_{O(n)},Z_2)$. Actually we could also define it using transgression, but in a fibering with fiber Z_2.

Proposition 22.4.

$$ H^*(B_{SO(n)},Z_2) = Z_2[w_2,\ldots,w_n], \quad d^o w_i = i. $$

This is an immediate consequence of theorem 18.3.

Our next objectives are to determine $H^*(B_{O(n)},Z_2)$ and to interpret the w_i as elementary symmetric functions. To do this we proceed analogously with the unitary case.

The ith Stiefel-Whitney class mod 2 of a principal bundle $(E,B,O(n))$ or $(E,B,SO(n))$ will then be defined as the image of the characteristic map of the corresponding universal class, or equivalently as the image by transgression of $x_{i-1}^{(n-i+1)}$ in the bundle $(E/O(i-1),B,V_{n,n-i+1})$ or $(E/SO(i-1),B,V_{n,n-i+1})$.

Let E be a principal bundle with fibre a finite group N, let Q be Abelian, $Q \subset N$, and let E be simply connected. Then N operates on the fibering $(E,E/Q,Q)$ by $xQ \to x\,n\,Q$, and hence N/Q operates on E/Q which is in fact a principal bundle $(E/Q,E/N,N/Q)$. For $y \in N/Q$ let $T_y : Q \to Q$ be defined by $q \to y^{-1}q\,y$ and let

$$y_* : \pi_1(E/Q) \to \pi_1(E/Q)$$

be the map induced by right translation of E/Q by y. Then there exists a canonical isomorphism

$$\phi : Q \to \pi_1(E/Q)$$

such that the following diagram is commutative:

$$\begin{array}{ccc} Q & \xrightarrow{\phi} & \pi_1(E/Q) \\ {\scriptstyle T_y}\downarrow & & \downarrow{\scriptstyle y_*} \\ Q & \xrightarrow{\phi} & \pi_1(E/Q) \end{array}$$

We define ϕ as follows. Let x be a fixed point of E/Q and let $\tilde{x}$ be any point of E over it. If $q \in Q$ we define $\phi_x(q)$ to be the element of $\pi_1(E/Q)$ whose representative loops are the projections of the paths in E which join $\tilde{x}$ to $\tilde{x}\cdot q$. Note that ϕ_x is independent of choice of $\tilde{x}$ since Q is Abelian.

Let $Q(n) \subset O(n)$ be the subgroup of diagonal matrices whose diagonal elements are ± 1, and let $SQ(n)$ denote the corresponding subgroup of $SO(n)$. Evidently

$$Q(n) \approx (Z_2)^n, \quad SQ(n) \approx (Z_2)^{n-1}.$$

Let $N_{Q(n)}$ be the normalizer of $Q(n)$ in $O(n)$ and define $X(O(n)) = N_{Q(n)}/Q(n)$. Then $X(O(n))$ operating on $Q(n)$ by inner automorphisms acts as $\sum n$, the symmetric group in n-variables. In what follows $Q(n)$ and $X(O(n))$ play the role of T^n and the Weyl group in the unitary case.

In the preceding discussion we now take $E = E_{O(n)}$, $Q = Q(n)$, and $N = N_{Q(n)}$; then $E_{O(n)}/Q(n) = B_{Q(n)}$ and $X(O(n))$ operates on $B_{Q(n)}$ naturally and on $\pi_1(B_{Q(n)})$ by maps induced by right translation. One sees readily that

$$H^*(B_{Q(n)},Z_2) = Z_2[u_1,\ldots,u_n], \; d^o u_i = 1.$$

In view of

$$H^1(B_{Q(n)},Z_2) = \mathrm{Hom}\,(\pi_1(B_{Q(n)}),Z_2)$$

and the above isomorphism ϕ it follows that $X(O(n))$ operates on $H^*(B_{Q(n)},Z_2)$ by permuting the generators u_i.

The fibre map

$$\rho(Q(n),O(n)) : B_{Q(n)} \to B_{O(n)}$$

and its induced map $\rho^*(Q(n),O(n))$ commute with the operations of $X(O(n))$. Since $X(O(n))$ clearly operates trivially on $H^*(B_{O(n)},Z_2)$ it follows that

$$\rho^*(Q(n), O(n))H^*(B_{O(n)},Z_2) \subset S(u_1,\ldots,u_n).$$

Similar remarks apply if consider $SO(n)$ and $SQ(n)$ in place of $O(n)$ and $Q(n)$. In particular, we note

$$H^*(B_{SQ(n)},Z_2) = Z_2[\bar{u}_1,\ldots,\bar{u}_{n-1}], \quad d^0\bar{u}_k = 1.$$

Note that the fibre in both cases is the same since $O(n)/Q(n) = SO(n)/SQ(n)$. We shall denote it by F_n.

<u>Lemma 22.5</u>. $\dim H^1(F_n,Z_2) \geq n-1$.

<u>Proof</u>. In the spectral sequence of $(B_{SQ(n)},B_{SO(n)},F_n)$ we have

$$E_2 \approx H^*(B_{SO(n)},Z_2) \otimes H^*(F_n,Z_2).$$

Since $B_{SO(n)}$ is simply connected $E_2^{1,0} = 0$; hence

$${}^1E_2 = E_2^{0,1} = H^1(F_n,Z_2),$$
$${}^1E_\infty = H^1(B_{SQ(n)},Z_2) = (Z_2)^{n-1}.$$

Then

$$\dim H^1(F_n,Z_2) = \dim {}^1E_2 \geq \dim {}^1E_\infty = n-1.$$

<u>Proposition 22.6</u>. $H^*(F_n,Z_2)$ is generated by elements of degree 1, and $P_2(F_n,t) = (1-t^2)(1-t^3)\cdots(1-t^n)(1-t)^{1-n}$.

<u>Proof</u>. We proceed by induction on N. For $n = 2$,

$$F_2 = SO(2)/Z_2 \approx T^1/Z_2 \approx T^1 = S^1$$

and hence the proposition holds. Assume the proposition for $n-1$, $(n > 2)$, and consider the inclusions

$$O(n) \supset Z_2 \times O(n-1) \supset Z_2 \times Q(n-1)$$

where $Z_2 \times O(n-1)$ and $Z_2 \times Q(n-1)$ are embedded in the usual way. This defines a fibering

$$(F_n,O(n)/Z_2 \times O(n-1),F_{n-1}) = (F_n,P_{n-1},F_{n-1})$$

where P_{n-1} is n-1 dimensional real projective space. In its spectral sequence we have

$$E_2 \approx H^*(P_{n-1},\mathcal{H}^*(F_{n-1},Z_2)),$$
$$E_2^{1,0} \approx Z_2, \quad E_2^{0,1} \approx H^1(F_{n-1},Z_2)^f.$$

Then

$$\dim {}^1E_2 = 1 + \dim H^1(F_{n-1},Z_2)^f \leq 1 + \dim H^1(F_{n-1},Z_2).$$

Using the inductive assumption the formula for $P_2(F_{n-1},t)$ gives

$$\dim H^1(F_{n-1},Z_2) = n-2;$$

hence $\dim {}^1E_2 \geq n-1$. On the other hand,

$$\dim {}^1E_2 \geq \dim {}^1E_\infty = \dim H^1(F_n,Z_2) \geq n-1$$

by the preceding lemma, and hence

$$n-1 \leq 1 + \dim H^1(F_{n-1},Z_2)^f \leq 1 + \dim H^1(F_{n-1},Z_2) \leq n-1.$$

Thus $H^1(F_{n-1},Z_2)^f = H^1(F_{n-1},Z_2)$ which shows that all the elements of $E_2^{0,1} \approx H^1(F_{n-1},Z_2)$ are permanent cocycles and that $\hat{H}^*(F_{n-1},Z_2)$ is simple in dimension 1. Since by our inductive assumption $H^*(F_{n-1},Z_2)$ is generated by elements of degree 1 we may conclude that $\hat{H}^*(F_{n-1},Z_2)$ is a simple system and all the elements of $H^*(F_{n-1},Z_2)$ are permanent cocycles. Thus

$$E_2 \approx H^*(P_{n-1},Z_2) \otimes H^*(F_{n-1},Z_2),$$

$dr \equiv 0$ for $r \geq 2$, and $E_2 = E_\infty$. Moreover since

$$H^*(P_{n-1},Z_2) = Z_2[x]/(x^n)$$

we have

$$P_2(F_n,t) = \frac{1-t^n}{1-t} \cdot P_2(F_{n-1},t)$$

which proves the desired formula. Note that each factor in $E_\infty = E_2$ is generated by elements of degree 1. Then by an elementary result on filtered rings the result follows for $H^*(F_n,Z_2)$, and the proposition is proved.

<u>Theorem 22.7</u>. $\rho^*(Q(n),O(n))$ maps $H^*(B_{O(n)},Z_2)$ isomorphically onto $S(u_1,\ldots,u_n)$ and maps w_i onto σ_i.

<u>Proof</u>. Consider the spectral sequence of the fibering $(B_{Q(n)},B_{O(n)},F_n)$:

$$E_2 \approx H^*(B_{O(n)},\hat{H}^*(F_nZ_2)),$$
$$E_2^{1,0} \approx H^1(B_{O(n)},Z_2), \quad E_2^{0,1} \approx H^1(F_n,Z_2)^f,$$
$$E_\infty = GrH^*(B_{Q(n)},Z_2).$$

Then

$$\dim {}^1E_2 \geq \dim {}^1E_\infty = \dim H^1(B_{Q(n)},Z_2) = n;$$
$$\dim {}^1E_2 = \dim E_2^{1,0} + \dim E_2^{0,1},$$
$$= 1 + \dim H^1(F_n,Z_2)^f ,$$
$$\leq 1+\dim H^1(F_n,Z_2) \leq 1 + (n-1) = n.$$

Therefore $\dim {}^1E_2 = n$ and

$$n = 1 + \dim H^1(F_n,Z_2)^f \leq 1 + \dim H^1(F_n,Z_2) = n,$$

from which it follows that $H^1(F_n,Z_2)^f = H^1(F_n,Z_2)$. Thus $E_2^{0,1} = E_\infty^{0,1}$ so that all the elements of $E_2^{0,1} \approx H^1(F_n,Z_2)$ are permanent cocycles and $\mathcal{H}^*(F_n,Z_2)$ is simple in dimension 1. In view of proposition 22.6 it follows that $\mathcal{H}^*(F_n,Z_2)$ is simple and $H^*(F_n,Z_2)$ consists of permanent cocycles. Then

$$E_2 \approx H^*(B_{O(n)},Z_2) \otimes H^*(F_n,Z_2),$$

$d_r \equiv 0$ for $r \geq 2$, and so $E_2 \approx E_\infty$. By theorems 14.1 and 14.2(a) we see that $\rho^*(Q(n),O(n))$ is injective. By 14.2(c) we have

$$P_2(B_{O(n)},t)\cdot P_2(F_n,t) = P_2(B_{Q(n)},t),$$

and hence

$$P_2(B_{O(n)},t) \cdot \frac{(1-t^2)\cdots(1-t^n)}{(1-t)^{n-1}} = \frac{1}{(1-t)^n},$$

$$P_2(B_{O(n)},t) = \frac{1}{(1-t)(1-t^2)\cdots(1-t^n)}.$$

But this is precisely $P_2(S(u_1,\ldots,u_n),t)$ which shows that $\rho^*(Q(n),O(n))$ maps $H^*(B_{O(n)},Z_2)$ onto $S(u_1,\ldots,u_n)$.

The proof of the second statement is analogous to the proof of proposition 21.6 and we leave it to the reader.

Corollary 22.8.

(a) $H^*(F_n,Z_2) = Z_2[u_1,\ldots,u_n]/(S^+(u_1,\ldots,u_n))$.

(b) $H^*(O(n)/Q(n),Z_2) = H^*(SO(n)/SQ(n),Z_2)$ is equal to its characteristic ring.

Proposition 22.9. $\rho^*(SO(n),O(n))$ is surjective and has (w_1) as kernel.

Proof. Consider the fibering $(B_{O(n)},B_{O(n)/SO(n)},B_{SO(n)},\pi)$. Note that $O(n)/SO(n) = Z_2$. We have

$$E_2 \approx H^*(B_{Z_2},\mathcal{H}^*(B_{SO(n)},Z_2)),$$
$$E_\infty = GrH^*(B_{O(n)},Z_2).$$

We also have

$$P_2(B_{Z_2},t) = 1/1-t,$$
$$P_2(B_{O(n)},t) = 1/(1-t)(1-t^2)\cdots(1-t^n),$$
$$P_2(B_{SO(n)},t) = 1/(1-t^2)\cdots(1-t^n);$$

the last formula is a consequence of proposition 22.4. Therefore

$$P_2(B_{Z_2},t) \cdot P_2(B_{SO(n)},t) = P_2(B_{O(n)},t),$$

and applying theorem 14.4 we have that $\mathcal{H}^*(B_{SO(n)},Z_2)$ is simple and i^* is surjective. By case III in section 17 we may identify i^* with $\rho^*(SO(n),O(n))$. It remains to determine the kernel. By 14.2(b) we have that i^* identifies

$$H^*(B_{O(n)},Z_2)/(\pi^* H^*_+(B_{Z_2},Z_2)) = H^*(B_{SO(n)},Z_2),$$
$$= Z_2[w_2,\ldots,w_n].$$

By an argument on dimension the left side is then readily seen to be $Z_2[w_1,\ldots,w_n]/(w_1)$

Remarks. (a) In $H^*(B_{Q(n)},Z_2)$ the Steenrod squares are given by

$$Sq^o u_i = u_i, \quad Sq^1 u_i = u_i^2$$

and by Cartan's product formula. It follows that

$$Sq^1(\sigma_j(u_1,\ldots,u_n)) = \sum u_1^2\cdots u_i^2 u_{i+1}\cdots u_j$$

where the right side denotes the symmetric function with the given summand as a typical term. Thus the determination of $Sq^1 w_j$ is reduced to the problem of expressing this symmetric function as a polynomial in the elementary symmetric functions with coefficients mod 2. We shall not give the solution, due to Wu Wen Tsün [8](see also [1]), but shall merely derive a formula to be used later. We have

$$Sq^1(\sigma_j(u_1,\ldots,u_n)) = \sum u_1^2 u_2\cdots u_j = \sigma_1\cdot\sigma_j-(j+1)\sigma_{j+1};$$

therefore

$$Sq^1 w_j = w_1 w_j + (j+1)w_{j+1}.$$

In $H^*(B_{SO(n)},Z_2)$ we then have

$$Sq^1 w_j = (j+1)w_{j+1}, \quad (2 \leq j \leq n).$$

(b) The identification of w_j with elementary symmetric functions allows us to derive a duality formula from the identity (21.9).

23. Orthogonal groups, cohomology mod $p \neq 2$.

For n = 2m and n = 2m+1 respectively a maximal torus T^m in SO(n) is given by the diagonal matrices

$$\begin{pmatrix} D_1 & & 0 \\ & \ddots & \\ 0 & & D_m \end{pmatrix} \qquad\qquad \begin{pmatrix} 1 & & & \\ & D_1 & & 0 \\ & & \ddots & \\ & 0 & & D_m \end{pmatrix}$$

where

$$D_i = \begin{pmatrix} \cos 2\pi x_i & \sin 2\pi x_i \\ -\sin 2\pi x_i & \cos 2\pi x_i \end{pmatrix}$$

The x_i are coordinates in the universal covering of T^m, and the preimage of the unit element of T^m is represented by points with integral coordinates. Hence we may identify the x_i with a basis of $H^1(B_{T^m},Z)$. The normalizer of T^m contains the permutations of the D_i; we may also replace D_i by AD_iA^{-1} where $A = \begin{pmatrix} 0 & 1 \\ 1 & 0 \end{pmatrix}$, thereby changing x_i into $-x_i$.

Since A has determinant equal to -1, for $n = 2m$ we must make an even number of such transformations. For n odd we can always obtain +1 as determinant by using -1 as first entry in the matrix. These transformations generate the full normalizer as may be easily seen, so the Weyl group $W(SO(2m+1))$ is the group of permutations of the x_i modulo an arbitrary number of sign changes, whereas $W(SO(2m))$ is the group of permutations modulo an even number of sign changes. T^m is also a maximal torus in $O(n)$, $(n = 2m, 2m+1)$, and analogously we see that $W(O(n))$ is the group of permutations of the x_i combined with an arbitrary number of changes in sign. Let y_i be images of the x_i by transgression; then

$$I_{SO(2m+1)} = I_{O(2m+1)} = I_{O(2m)} = S(y_1^2,\dots,y_m^2),$$

and $I_{SO(2m)}$ is the ring generated by the elements

$$\sigma_i(y_1^2,\dots,y_m^2), \quad (1 \leq i \leq m-1),$$

and the product $y_1 \cdots y_m$. To prove the latter it suffices to show that if in a polynomial $P(y_1,\dots,y_m)$ which is invariant under $W(SO(2m))$ a monomial occurs which has at least one odd exponent then it is divisible by $y_1 \cdots y_m$. This can be easily shown using the operations of $W(SO(2m))$ and even numbers of sign changes.

Proposition 23.1. For $p \neq 2$, $SO(n)$ has no p-torsion, and

(a) $H^*(SO(2m+1),Z_p) = \Lambda(x_3,x_7,\dots,x_{4m-1})$,

(b) $H^*(SO(2m),Z_p) = \Lambda(x_3,x_7,\dots,x_{4m-5},x_{2m-1})$.

Proof. (a) may be proved readily by induction making use of the spectral sequence of the fibering

$$SO(2m+1)/SO(2m-1) = V_{2m+1,2}\ ,$$

and 22.1(b).

(b) Consider the spectral sequence of the fibering

$$SO(2m)/SO(2m-1) = S_{2m-1}\ ;$$

we have

$$E_2 = \Lambda(x_{2m-1}) \otimes \Lambda(x_3,x_7,\dots,x_{4m-5}),$$

and it suffices to show that $E_2 = E_\infty$. Since the only non-trivial base degrees are 0 and $2m-1$, only d_{2m-1} may not be identically zero, and hence $E_{2m-1} = E_2$ and $E_{2m} = E_\infty$. On the other hand the x_i are universally transgressive and have odd degrees; hence $d_{2m-1}x_i = 0$, whence $d_{2m-1} = 0$, $E_2 = E_\infty$. Since

$$P_p(SO(n),t) = P_o(SO(n),t), \quad (p \neq 2),$$

this shows further that $SO(n)$ has no p-torsion.

Theorem 23.2. If $p \neq 2$ then

(a) $\rho^*(T^m, SO(2m+1))$ maps $H^*(B_{SO(2m+1)}, Z_p)$ isomorphically onto $S(y_1^2, \ldots, y_m^2)$,

(b) $\rho^*(T^m, SO(2m))$ maps $H^*(B_{SO(2m)}, Z_p)$ isomorphically onto the ring generated by $S(y_1^2, \ldots, y_m^2)$ and the product $y_1 \cdots y_m$.

(c) $\rho^*(T^m, O(n))$ maps $H^*(B_{O(n)}, Z_p)$ isomorphically onto $S(y_1^2, \ldots, y_m^2)$ for $n = 2m$, $2m+1$.

The theorem is a consequence of 18.1, 20.3, the remark following 20.4, 23.1, and the initial remarks of this section.

Corollary 23.3. If $p \neq 2$ then $\rho^*(SO(n), O(n))$ is injective; $B_{SO(n)}$ and $B_{O(n)}$ have no p-torsion.

24. Integral cohomology of $B_{O(n)}$ and $B_{SO(n)}$.

Let X be a space with finitely generated integral cohomology groups. By the universal coefficient theorem

$$H^i(X, Z_2) = H^i(X, Z) \otimes Z_2 + \mathrm{Tor}(H^{i+1}(X, Z), Z_2).$$

Assume that the 2-primary component of $H^i(X, Z)$ is a direct sum of q_i cyclic groups, and that the dimension of $H^i(X, Z_o)$ is p_i; then

$$P_2(X, t) = \sum (p_i + q_i + q_{i+1}) \cdot t^i = P_o(X, t) + (1 + 1/t) \sum q_i \cdot t^i.$$

As is well known, Sq^1 is the Bockstein homomorphism attached to the exact sequence

$$0 \to Z \to Z \to Z_2 \to 0$$

followed by reduction mod 2. From this one deduces readily:

Lemma 24.1. If X has finitely generated integral cohomology and A is the graded subspace $Sq^1(H^*(X, Z_2))$ then the 2-primary component of $H^*(X, Z)$ consists only of elements of order 2 if and only if

$$P_2(X, t) - P_o(X, t) = (1 + 1/t)\, P(A, t).$$

Lemma 24.2. If X has finitely generated integral cohomology groups whose torsion subgroups are direct sums of cyclic groups of order 2 then an element $x \in H^*(X, Z)$ is completely determined by its images x_o and x_2 in $H^*(X, Z_o)$ and $H^*(X, Z_2)$ respectively.

Proof. We have to show that if $x_o = x_2 = 0$ then $x = 0$. If $x_o = 0$ then clearly x is a torsion element and therefore has order 2. But then $x_2 = 0$ implies $x = 0$.

We now consider a commutative graded algebra of finite type with $H^0 = K$. Let D be a derivation of H of degree +1 and such that $D^2 = 0$, (for example, Sq^1 on $H^*(X, Z_2)$). We introduce the following notation:

A : graded subspace stable under D,

N_A : the kernel of D in A,

M_A : a supplementary subspace to N_A,

I_A : the image of A under D,

J_A : a supplementary subspace to I_A (in A).

Now let B be a second subspace stable under D and linearly disjoint from A over K--i.e., the map $a \otimes b \to a \cdot b$ of $A \otimes B \to H$ is injective. We denote the image of this map by A.B.

We propose to show that

(24.3) $$P(J_{A \cdot B},t) = P(J_A,t) \cdot P(J_B,t).$$

Since $A = M_A + I_A + J_A$, and since D is an isomorphism of M_A onto I_A, we have

$$P(A,t) = (1 + 1/t)P(I_A,t) + P(J_A,t),$$

and analogously,

(24.4) $$P(B,t) = (1 + 1/t)P(I_B,t) + P(J_B,t),$$

$$P(A \cdot B,t) = (1 + 1/t)P(I_{A \cdot B},t) + P(J_{A \cdot B},t).$$

Since A and B are linearly disjoint we have

$$P(A \cdot B,t) = P(A,t) \cdot P(B,t),$$

from which we obtain

(24.5) $$P(A \cdot B,t) = (1 + 1/t)^2 P(I_A,t)P(I_B,t) + (1 + 1/t)(P(I_A,t)P(J_B,t) + P(J_A,t)P(I_B,t)) + P(J_A,t)\ P(J_B,t).$$

On the other hand the image of D is spanned by $I_A(I_B + J_B)$, $I_B(I_A + J_A)$, and $D(M_A \cdot M_B)$. On $M_A \cdot M_B$ the derivation D is clearly injective; hence

$$P(D(M_A \cdot M_B),t) = tP(M_A,t)P(M_B,t),$$
$$= 1/t\ P(I_A,t)P(I_B,t).$$

Making use of this one sees readily that $(1+1/t)\ P(D(A \cdot B),t)$ is equal to the sum of the first three terms on the right side of 24.5; hence 24.3 follows from 24.4.

Theorem 24.6. The torsion elements of $H^*(B_{SO(n)},Z)$ are of order 2.

Proof. We have seen that $B_{SO(n)}$ has no p-torsion for $p \neq 2$, and that

$$H^*(B_{SO(n)}, Z_2) = Z_2|w_2,\dots,w_n|$$

where $Sq^1 w_i = (i-1)w_{i+1}$. Therefore

$$H^*(B_{SO(2m+1)},Z_2) \simeq A_1 \otimes \cdots \otimes A_m$$

where $A_i = Z_2[w_{2i},w_{2i+1}]$ and $A_1,\dots,A_m$ are stable under Sq^1 and the cupproduct; and

$$H^*(B_{SO(2m)},Z_2) \simeq A_1 \otimes \cdots \otimes A_m$$

where $A_1,\ldots,A_{m-1}$ are as before and $A_m = Z_2[w_{2m}]$ is annihilated by Sq^1. In A_i the image of Sq^1 is spanned by the elements $w_{2i}^s \cdot w_{2i+1}^t$, ($s \geq 0$, s even, $t > 0$):consequently we may take the space spanned by the elements $w_{2i}^s \cdot w_{2i+1}^t$, (s odd, $t > 0$), as M_{A_i}. Similarly for J_{A_i} we may take the space spanned by w_{2i}^s, (s even). It follows that

$$P(J_{A_i},t) = \begin{cases} (1-t^{4i})^{-1}, & \text{if } i \neq m \text{ or } n \neq 2m, \\ P(J_{A_m},t) = P(A_m,t) = (1-t^{2m})^{-1}, & \text{if } i = m,\ n = 2m. \end{cases}$$

Applying (24.3) to m factors we get for $n = 2m+1$,

$$P_2(B_{SO(n)},t) = (1 + 1/t)\, P(Sq^1(H^*(B_{SO(n)},Z_2),t) + \Pi_{i=1}^m (1-t^{4i})^{-1};$$

and for $n = 2m$,

$$P_2(B_{SO(n)},t) = (1 + 1/t)\, P(Sq^1(H^*(B_{SO(n)},Z_2),t) + (1-t)^{2m-1}\Pi_{i=1}^m (1-t^{4i})^{-1}$$

Since in both cases the last term is $P_o(B_{SO(n)},t)$ by 23.2, the theorem follows from lemma 24.1.

Theorem 24.7. The torsion elements of $H^*(B_{O(n)},Z)$ are of order 2.

Proof. We have

$$H^*(B_{O(n)},Z_2) \approx Z_2[w_1,\ldots,w_n],$$
$$Sq^1 w_i = w_1 w_i + (i-1)w_{i+1}.$$

We choose a new basis

$$w_1^* = w_1, w_{2i}^* = w_{2i},\ w_{2i+1}^* = w_{2i+1} + w_{2i}\cdot w_1;$$

then

$$H^*(B_{O(n)},Z_2) \approx Z_2[w_1^*,\ldots,w_n^*],$$
$$Sq^1 w_1^* = (w_1^*)^2,$$
$$Sq^1 w_{2i}^* = w_{2i+1}^*,$$
$$Sq^1 w_{2i+1}^* = Sq^1 w_{2i+1} + w_1^2 \cdot Sq^1 w_{2i} + w_1(w_1 w_{2i} + w_{2i+1}) = 0.$$

This gives

$$H^*(B_{O(2m+1)},Z_2) = A_0 \otimes A_1 \otimes\cdots\otimes A_m$$

where

$$A_0 = Z_2[w_1^*],\ A_i = Z_2[w_{2i}^*,w_{2i+1}^*],$$

and $A_o,\ldots,A_m$ are stable under Sq^1 ; and

$$H^*(B_{O(2m)},Z_2) = A_0 \otimes A_1 \otimes\cdots\otimes A_{m-1}$$

where

$$A_0 = Z_2[w_1^*,w_{2m}^*],\ A_i = Z_2[w_{2i}^*,w_{2i+1}^*]$$

and $A_0,\ldots,A_{m-1}$ are stable under Sq^1. As in the preceding theorem we have

$$P(J_{A_i},t) = (1-t^{4i})^{-1},\ (i > 0).$$

For n odd it is clear that the elements of strictly positive degrees in A_0 in the kernel of Sq^1 are w_{2s}^* and that they span $Sq^1(A_0)$; hence $P(J_A,t) = 1$. We assert that this holds also in case n is even. We have

$$Sq^1(w_1^s w_{2m}^t) = (s+t)w_1^{s+1}w_{2m}^t$$

which is zero if and only if s+t is even. But then for $s > 0$ it is equal to $Sq^1(w_1^{s-1}w_{2m}^t)$, and for $s = 0$ and t even, $t > 0$, it is equal to $Sq^1(w_{2m}^{t-1})$ from which we get $P(J_{A_0},t) = 1$. Now the remainder of the proof is the same as the proof in 24.6.

Corollary 24.8. The kernel of Sq^1 in $H^*(B_{SO(n)},Z_2)$, (and in $H^*(B_{O(n)},Z_2)$), consists of the integral cohomology reduced mod 2.

This follows from 24.1, 24.6, 24.7.

25. Stiefel-Whitney classes, Pontrjagin classes.

In view of 24.2 and 24.8 we see that there is a unique element of order 2 in $H^{2i+1}(B_{SO(n)},Z)$, (or in $H^{2i+1}(B_{O(n)},Z)$), whose reduction mod 2 is w_{2i+1}; we denote it by W_{2i+1}. Similarly W_2 is the element of order 2 in $H^2(B_{O(n)},Z)$ whose reduction mod 2 is w_1^2, (for $n \geq 3$). W_2 and W_{2i+1} (in the case of SO(n)) are the universal integral Stiefel-Whitney classes in the indicated dimensions. (In the case of O(n) we do not get the Stiefel-Whitney classes; these are defined with respect to "twisted" integral coefficients - we shall not discuss them.)

The element W_{2m} which is the image by transgression of a generator of $H^{2m-1}(S_{2m-1},Z)$ in the fibering

$$(B_{SO(2m-1)},B_{SO(2m)},S_{2m-1})$$

is called the universal Euler-Poincaré class. We recall that if ϕ is the classifying map of the unit tangent bundle of a differentiable manifold B of dimension 2m then

$$\phi^*(W_{2m}) = \chi(B)\cdot F$$

where $\chi(B)$ is the Euler characteristic and F the fundamental class of B. $\phi^*(W_{2m})$ is also called the 2m-Stiefel-Whitney class and in fact, by definition, its reduction mod 2 is the 2m-Stiefel-Whitney class mod 2. However, it is an element of infinite order and not of order 2 as are the other integral Stiefel-Whitney classes. The Stiefel-Whitney classes of a bundle are defined as usual by means of the characteristic map of the bundle.

Proposition 25.1. Let T^m be the maximal torus of SO(2m) described in section 23. Then

$$\rho^*(T^m,SO(2m))(W_{2m}) = y_1 \cdots y_m.$$

Proof. Consider the inclusion $i : U(m) \subset SO(2m)$. We have

$$U(m) \cap SO(2m-1) = U(m-1),$$

and i induces the identity map of $S_{2m-1} = U(m)/U(m-1)$ onto $S_{2m-1} = SO(2m)/SO(2m-1)$. We also have a commutative diagram

$$\begin{array}{ccc} B_{SO(2m-1)} & \xleftarrow{\alpha} & B_{U(m-1)} \\ \downarrow & & \downarrow \\ B_{SO(2m)} & \xleftarrow{\beta} & B_{U(m)} \end{array}$$

where all the maps are the corresponding $\rho(U,G)$ maps. Then α defines a homomorphism

$$(B_{U(m-1)}, B_{U(m)}, S_{2m-1}) \to (B_{SO(2m-1)}, B_{SO(2m)}, S_{2m-1})$$

from which it follows that $\beta^*(W_{2m})$ is the image by transgression of a generator $H^{2m-1}(S_{2m-1},Z)$--in other words $\beta^*(W_{2m}) = C_{2m}$ by definition of the Chern classes. Now T^m is also a maximal torus of $U(m)$; hence

$$\begin{aligned} \rho^*(T^m,SO(2m)) &= \rho^*(T^m,U(m))\cdot\rho^*(U(m),SO(2m)), \\ &= \rho^*(T^m,U(m))\cdot\beta^*, \end{aligned}$$

and the proposition follows from results in section 21

In general the integral Stiefel-Whitney classes do not obey duality, but the highest dimensional ones do.

Proposition 25.2. Let $(E^{(i)},B,SO(n_i))$, $(i = 1,2)$, be bundles and let $(E,B,SO(n))$ be their sum bundle. If $W_j^{(1)}, W_j^{(2)}$, and W_j denote their respective classes then

$$W_n = W_{n_1}^{(1)} \cdot W_{n_2}^{(2)} .$$

Proof. In view of the final remark of section 22 the formula to be proved holds when reduced mod 2. By 24.2 it thus suffices to consider rational coefficients. If n is odd then necessarily one of the n_i is odd, say n_1. Then the rational reductions of both W_n and $W_n^{(1)}$, are zero so that the formula holds. If n_1 and n_2 are both even then the formula follows from the identifications of W_{n_1} and W_{n_2} with the elementary symmetric functions σ_{n_1} and σ_{n_2}. Finally suppose both n_i are odd, it is sufficient to prove

$$\rho^*(SO(n_1) \times SO(n_2), SO(n))\,(W_n) = 0.$$

Let T be a maximal torus of $SO(n)$ chosen as in section 23, and let T' be a maximal torus of $SO(n_1) \times SO(n_2)$. We may choose T' so that $T' \subset T$ with $x_1 = 0$, (x_i arbitrary). Then in particular $\rho^*(T',T)\,(y_1) = 0$ and hence

$$\rho^*(T',T)(y_1,\ldots,y_n) = \rho^*(T',SO(n))(W_n) = 0.$$

But

$$\rho^*(T',SO(n)) = \rho^*(T',SO(n_1) \times SO(n_2))\cdot\rho^*(SO(n_1)\times SO(n_2),SO(n))$$

and $\rho^*(T',SO(n_1) \times SO(n_2))$ is injective over the rationals, so the assertion is proved.

As an application of 25.2 we cite the following theorem due to H. Samelson [6]:

Proposition 25.3. If M is a compact orientable manifold of dimension 2m which has a continuous field of odd dimensional plane elements then $\chi(M) = 0$.

The assumptions imply that the tangent bundle to M is a Whitney sum bundle of two bundles with groups $SO(2m_i+1)$, $(i = 1,2)$, $m_1+m_2+1 = m$. Therefore by 25.2, the 2m-dimensional Stiefel-Whitney class of M is a product of the corresponding highest dimensional classes of the two bundles and hence must be an element of order 2. But as noted earlier it is also equal to $\chi(M)\cdot F$, and since the fundamental class F has infinite order we must have $\chi(M) = 0$.

Let $SO(n) \subset O(n) \subset U(n)$ be the natural inclusions. The image of the Chern class C_{2i} under $\rho^*(SO(n),U(n))$, (or $\rho^*(O(n),U(n))$), prefixed by $(-1)^{i/2}$ when i is even, is called the Pontrjagin class in dimension 2i; We denote it by P_{2i}. Usually one considers only the classes P_{4i} : the classes P_{4i+2} which we introduce here for convenience are elements of order 2 as we shall see.

Let T^n be a maximal torus in U(n) and $v_1,\ldots,v_n$ a basis of $H^2(B_{T^n},Z)$ as described in section 22. The maximal tori T^m of SO(n) and O(n) have dimension $m = [n/2]$. We may without essential change replace O(n) by a conjugate subgroup in U(n), and it is convenient to choose it so that the corresponding tori T^m are now given by diagonal matrices of the form:

$$\begin{pmatrix} e^{2\pi i x_1} & & & & \\ & e^{-2\pi i x_1} & & & 0 \\ 0 & & \ddots & & \\ & & & e^{2\pi i x_m} & \\ & & & & e^{-2\pi i x_m} \end{pmatrix}, \quad \begin{pmatrix} 1 & e^{2\pi i x_1} & & & \\ & & e^{-2\pi x_1} & & \\ & & & \ddots & \\ & & & & e^{2\pi i x_m} \\ & & & & & e^{-2\pi i x_m} \end{pmatrix},$$

for n even and odd respectively. If we denote by $y_1,\ldots,y_m$ the basis of $H^2(B_{T^m},Z)$ where y_i is obtained from x_i by transgression then

$$\rho^*(T^m,T^n)(v_1) = 0, \ (n \text{ odd}),$$

$$\rho^*(T^m,T^n)(v_{2i}) = -\rho^*(T^m,T^n)(v_{2i+1}) = y_i, \ (n \text{ odd}),$$

$$\rho^*(T^m,T^n)(v_{2i-1}) = -\rho^*(T^m,T^n)(v_{2i}) = y_i, \ (1 \le i \le m), \ (n \text{ even}).$$

Therefore

$$\rho^*(T^m,T^n)\sigma_i(v_1,\dots,v_n) = 0, \text{ (i odd)},$$

$$\rho^*(T^m,T^n)\sigma_{2i}(v_1,\dots,v_n) = (-1)^i\sigma_i(y_1^2,\dots,y_m^2).$$

Now making use of the identities

$$\rho^*(T^m,T^n)\ \rho^*(T^n,U(n)) = \rho^*(T^m,O(n))\ \rho^*(O(n),U(n)),$$

$$\rho^*(T^m,T^n)\ \rho^*(T^n,U(n)) = \rho^*(T^m,SO(n))\ \rho^*(SO(n),U(n)),$$

we obtain the following:

Proposition 25.4. With the above notations and for integer coefficients we have

$$\rho^*(T^m,O(n))(P_{4i+2}) = 0,$$

$$\rho^*(T^m,O(n))(P_{4i}) = \sigma_i(y_1^2,\dots,y_m^2).$$

Clearly, the same formulas will hold for the reduction mod p of the Pontrjagin classes, and will characterise them for $p \neq 2$. Before considering the Pontrjagin classes reduced mod 2 we make the following observations. If T^n is a torus and $Q(n)$ its subgroup consisting of elements of order 2 then we recall

$$H^*(B_{T^n},Z_2) = Z_2[v_1,\dots,v_n], \quad (d^o v_i = 2),$$

$$H^*(B_{Q(n)},Z_2) = Z_2[u_1,\dots,u_n], \quad (d^o u_i = 1).$$

We now claim that

(25.5) $$\rho^*(Q(n),T^n)(v_i) = u_i^2,$$

(for the natural bases chosen as in sections 21 and 22). It clearly suffices to consider the case n = 1. Then $\rho^*(Q(1),T^1)$ is the projection map in the fibering $(B_{Q(1)},B_{T^1},T^1/Q(1))$ whose fibre is S_1. In the spectral sequence the term

$$E_2 = Z_2[v_1] \otimes H^*(S_1,Z_2)$$

has the same Poincaré polynomial as E_∞ and hence $E_\infty = E_2$. From this (25.5) follows readily.

Proposition 25.6.

(a) $P_{4i+2} = (W_{2i+1})^2$,

(b) $w_i^2 = P_{2i} \mod 2$.

Proof. It is sufficient to prove the proposition in the case of O(n) -- it will then follow for SO(n) upon application of $\rho^*(SO(n),O(n))$. From the definition of P_{2i}, (disregarding signs since we compute mod 2),

$$\begin{aligned}\rho^*(Q(n),O(n))(P_{2i}) &= \rho^*(Q(n),U(n))(C_{2i}),\\ &= \rho^*(Q(n),T^n)\ \rho^*(T^n,U(n))(C_{2i}),\\ &= \rho^*(Q(n),T^n)\ \sigma_i(v_1,\dots,v_n),\\ &= \sigma_i(u_1^2,\dots,u_n^2),\\ &= \rho^*(Q(n),O(n))(w_i^2),\end{aligned}$$

the last three steps following by 21.6, 25.5, and 22.7 respectively. Since $\rho^*(Q(n),O(n))$ is injective, (as noted in the proof of 22.7), (b) follows. Moreover we see that P_{4i+2} and W_{2i+1}^2 are equal when reduced mod 2. From 25.4 it follows that the rational reduction of P_{4i+2} is zero and hence equal to the rational reduction of $(W_{2i+1})^2$. Thus (a) follows because of 24.2.

Remarks. (a) In view of the results of section 24 the integral Pontrjagin classes are completely characterized by propositions 25.4 and 25.6.

(b) The integral classes P_{4i} by themselves do not obey duality (since obviously their mod 2 reductions do not); however taken together with the classes W_{2i+1}^2 they do. By 24.2 it suffices to prove this over the integers mod 2 and the rationals. In both cases the classes may be represented by elementary symmetric functions in the squares of certain variables; the duality relations then follow from an identity analogous to 21.9.

(c) Let G be a connected compact Lie group. The Weyl group W(G) in operating on the fibering $(B_T,B_G,G/T)$ of course acts trivially on B_G; hence $\rho^*(T,G) \subset I_G$. We recall from section 2o. that the kernel of $\rho^*(T,G)$ consists of the torsion subgroup of $H^*(B_G,Z)$. For the case $G = U(n)$ the image of $\rho^*(T,G)$ is I_G, and in fact the same is true for any group without torsion. The results of section 24 and propositions 25.1 and 25.4 show that it is also true for $G = SO(n)$ and $G = O(n)$. It is not known whether it is true in general.

(d) The integral Stiefel-Whitney and Pontrjagin classes may also be defined by transgression. Consider a bundle

$$(E,B,SO(n)/SO(m)),\ m = 2s-1\ .$$

Then, (see [2] section 1o), the lowest dimensional non-vanishing groups of the fibre are

$$H^0(V_{n,n-m},Z) = H^{4s-1}(V_{n,n-m},Z) = Z$$
$$H^{2s}(V_{n,n-m},Z) = Z_2\ .$$

Then W_{2s+1} is the image by transgression of the generator of $H^{2s}(V_{n,n-m},Z)$. A generator $x \in H^{4s-1}$ is not in general transgressive; but 2x is always transgressive and its image is then the Pontrjagin class.

BIBLIOGRAPHY

[1] A. Borel, La cohomologie mod 2 de certains espaces homogènes, Comm. Math. Helv. 27 (1953), 165-197.

[2] A. Borel, Ann. of Math. 57 (1953), 115-2o7.

[3] A. Borel, J.P. Serre, Am. J. Math. 75(1953), 4o9-448.

[4] S.S. Chern, On the characteristic classes of complex sphere bundles and algebraic varieties, Amer. J. Math. 75 (1953), 565-597.

[5] L. Pontrjagin, Characteristic cycles on differentiable manifolds, Mat. Sbornik N.S. 21, 63 (1947), 233-284.

[6] H. Samelson, A theorem on differentiable manifolds, Port. Math. 1o (1951), 129-133.

[7] N. Steenrod, The topology of fibre bundles, Princeton U. Press (1951).

[8] Wu Wen Tsün, Les i-carrés dans une variété grassmannienne, C.R.Acad.Sci. Paris 23o (195o), 918-92o.

BIBLIOGRAPHICAL NOTES AND COMMENTS

(added in 1967)

CHAPTER I

The standard reference for Hopf algebras is now [12].

CHAPTER II

There are now a number of expositions of spectral sequences, see e.g. [4], or, for a treatment geared to the discussion of fibre bundles, [7].

CHAPTER III

For a more general construction of universal bundles, we refer to [10].The transgression Theorem 18.1, stated without proof here, can be proved more easily than in reference [1] of Chapter III, to be quoted hereafter by {1}, by use of the Eilenberg-Moore spectral sequence [13]. See also an announcement [14] and forthcoming papers by Rothenberg-Steenrod, where a similar spectral sequence is introduced in a more geometric context. Strictly speaking, these results do not include the purely algebraic theorem of {1}, but they cover the topological applications, and yield further results as well. In § 19, we have proved a special case of 18.1 by a method similar to the method of {1}, in order to give an idea of the general proof. However, this special case can be handled more simply by means of the comparison theorem of spectral sequences [16, 17]. The basic tool in the proof of Theorems 18.1 and 19.1 is the notion of a set of elements in a ring which are "annihilator-free" (or "without relations" in the terminology of {1}), suggested in part by earlier work of Leray and of Koszul. A similar concept was introduced in 1955 by Serre in the theory of local rings, under the name of E-sequence, and this terminology has now become customary in homological algebra It occurs notably in dimension theory, and the starting point is a lemma quite similar to 19.2. For this, see [15], where further references are also given.

The transgression theorems mentioned in § 18 imply for a compact connected Lie group G and a prime p : (1) If G has no p-torsion, then its classifying space B_G has no p-torsion; (2) If B_G has no p-torsion and $H^*(B_G,K_p)$ is a polynomial ring, then G has no p-torsion. In [1] it is checked, case by case, that if B_G has no p-torsion, then G has no p-torsion. To prove this _a priori_, using (2), one needs to know that the first assumption of (2) implies the second one. This implication can be deduced from a result of Serre's, which states that if a polynomial algebra is a free module of finite rank over a sub-algebra, then the latter is itself a polynomial algebra. In fact, assume that B_G has no p-torsion and let T be a maximal torus of G. Then, in view of known facts about B_G and G/T, the E_2-term of the spectral sequence in cohomology mod p of the fibering $(B_T, B_G, G/T)$ contains non-zero terms only in even degrees, hence $E_2 = E_\infty = H^*(B_G, K_p) \otimes H^*(G/T, K_p)$ which implies that $H^*(B_T, K_p)$ is a free module of finite rank over $H^*(B_G, K_p)$. Since the former is a polynomial algebra, Serre's result implies that the same is true for $H^*(B_G, K_p)$.

The p's for which G has no p-torsion have been completely determined and this leads to interesting relationships between various properties of G, which have been partly checked, partly proved a priori (see [1], [6]; see also Theorem 4.2 of [4]).

Bott's paper [3] gives the proofs of the results announced in reference [4] of Chapter III.

CHAPTER IV

The characteristic classes for the classical groups are introduced here in the framework of the cohomology of classifying spaces, invariants of the Weyl group, etc. This approach is convenient to compute characteristic classes of homogeneous vector bundles, or of bundles associated to bundles of frames on a manifold (see [2]), but presupposes a certain amount of machinery. For a more geometric treatment, based on a direct study of Grassmannians, we refer to Milnor's Lecture Notes [11]. Chern classes have also been given an axiomatic description by Hirzebruch [8], which has been carried over in algebraic geometry by Grothendieck [5]. An exposition along those lines is given in [9]. For a comparison between the different definitions of Chern classes, see [2, Appendix I]. Much of §§ 24, 25 has been incorporated in [2, Appendix II].

REFERENCES

[These are the references for the Appendix. Each chapter carries its own bibliography.]

[1] A. Borel, Sous-groupes commutatifs et torsion des groupes de Lie compacts connexes, Tohoku Math. J. 13 (1961), 216-240.

[2] A. Borel and F. Hirzebruch, Characteristic classes and homogeneous spaces I, Amer. Jour. Math. 80 (1958), 459-538, II, ibid. 81 (1959), 315-382, III, ibid. 82 (1960), 491-504.

[3] R. Bott, An application of the Morse theory to the topology of Lie groups, Bull. Soc. Math. France 84 (1956), 251-282.

[4] H. Cartan and S. Eilenberg, Homological Algebra, Princeton University Press, Princeton, 1956.

[5] A. Grothendieck, La théorie des classes de Chern, Bull. Soc. Math. France 86 (1958), 137-154.

[6] B. Harris, Torsion in Lie groups and related spaces, Topology 5 (1966), 347-354.

[7] P. J. Hilton and S. Wylie, Homology Theory, Cambridge University Press, Cambridge, 1960.

[8] F. Hirzebruch, Topological Methods in Algebraic Geometry, 3rd ed., Springer-Verlag, Berlin, 1966.

[9] D. Husemoller, Fibre Bundles, McGraw-Hill, New York, 1966.

[10] J. Milnor, Construction of universal bundles I, II, Annals of Math. (2) 63 (1956), 272, 430-436.

[11] J. Milnor, Lectures on characteristic classes, Princeton University 1967, mimeographed. (Notes by J. Stasheff.)

[12] J. Milnor and J. C. Moore, On the structure of Hopf algebras, Annals of Math. (2) 81 (1965), 211-264.

[13] J. Moore, Algèbre homologique et homologie des espaces classifiants, Sém.E.N.S. 1959, Exposé 7.

[14] M. Rothenberg and N. Steenrod, The cohomology of classifying spaces of H-spaces, Bull. Amer. Math. Soc. 71 (1965), 872-875.

[15] J. P. Serre, Algèbre locale. Multiplicités, (rédigé par P. Gabriel), Lecture Notes in Mathematics 11 (1965), Springer.

[16] E. C. Zeeman, A proof of the comparison theorem for spectral sequences, Proc. Cambridge Phil. Soc. 53 (1957), 57-62.

[17] E. C. Zeeman, A note on a theorem of Armand Borel, ibid. 396-398.